Radiation Protection in Nuclear Medicine

Sören Mattsson • Christoph Hoeschen
Editors

Radiation Protection in Nuclear Medicine

Springer

Editors
Sören Mattsson
Medical Radiation Physics
Department of Clinical Sciences Malmö
Lund University
Skåne University Hospital Malmö
SE-205 02 Malmö
Sweden

Christoph Hoeschen
Helmholtz Zentrum München
Deutsches Forschungszentrum für
Gesundheit und Umwelt (GmbH)
Ingolstädter Landstr. 1
DE-85764 Neuherberg
Germany

ISBN 978-3-642-43716-8 ISBN 978-3-642-31167-3 (eBook)
DOI 10.1007/978-3-642-31167-3
Springer Heidelberg New York Dordrecht London

Softcover reprint of the hardcover 1st edition 2013

Printed on acid-free paper

Springer is part of Springer Science+Business Media (www.springer.com)

Preface

Radiation protection is an important task for nuclear medicine departments with regard to the use of radiopharmaceuticals for diagnostic and therapeutic purposes. It is important to make sure that the radiation dose to the patients as well as to staff members is kept "as low as reasonable achievable" (ALARA). In the last few years, the use of PET substances with higher photon energy than the previously used radionuclides has been more and more common, and the number of PET radionuclides is increasing. There is a sudden appearance of multimodal PET/CT and SPECT/CT scanners. New radiopharmaceuticals for targeted radionuclide therapy are introduced. Radiotracers are more and more used in surgical practices like identification of lymph node involvement in breast cancer and colon cancer, etc. The staff will be exposed to radiation during production, labeling, transport, injection, and when being close to the patients. The increased use of PET-imaging causes a need for new planning of radiation protection and education of all categories of staff members. This was the reason for choosing the topic of radiation protection in nuclear medicine for the second training course organized by the MADEIRA (Minimizing Activities and Doses by Enhancing Image quality in Radiopharmaceutical Administration) project, cofunded by the European Commission through the EURATOM Seventh Framework Programme.

This book is the second book of a series of three corresponding to such training courses. It presents articles related to quantities and units as well as basic radiobiology for radiation protection. There is a special chapter about the radiobiology and dosimetry for the lens of the eye and another one about the protection of embryo and fetuses. As the quality of the equipment influences the image quality as well as the patient dose, there is a specific chapter about QC of gamma cameras and SPECT/CT and PET/CT units. Measurements and calculations of doses are covered and examples of shielding calculations for PET/CT installations are given. Releases to the environment through releases from laboratories as well releases through patient excreta are discussed. Finally there is a chapter on "Rules of the thumb" for radiation protection in nuclear medicine.

The book is aimed for medical physicists, technicians, physicians in nuclear medicine and radiology, radiochemists, engineers, PhD students, radiation protection experts, and others involved in nuclear medicine, radionuclide production, and radiation protection.

Sören Mattsson
Christoph Hoeschen

Contents

Chapter 1
Introduction: The Importance of Radiation Protection in Nuclear Medicine

Sören Mattsson

1.1 The Importance of Radiation Protection in Nuclear Medicine

In nuclear medicine, radiopharmaceuticals are administered to the patient either for the production of diagnostic images (diagnostic nuclear medicine or molecular imaging) or with the intention to treat using the emitted radiation from the radiopharmaceutical (nuclear medicine therapy). The most common way for administration is through an intravenous injection. The radiopharmaceutical is sometimes swallowed by the patient. Alternatively, the patient may breathe a radioactive gas or aerosol.

Impressive progresses have taken place within diagnostic nuclear medicine during the last few years. Diagnostic procedures are now more and more performed using PET/CT and SPECT/CT units. Especially the PET units require specific site planning and shielding.

In radionuclide therapy, still dominated by radioiodine therapy for thyreotoxicosis and thyroid cancer, there is also an increasing use of radionuclide-labeled monoclonal antibodies and peptides. At therapy, the activities are higher, and the radionuclides used are often different from those used in diagnostic nuclear medicine. They are usually beta emitters (sometimes also low-energy electron or alpha emitters) with longer physical and biological half-lives and therefore constitute a greater radiation protection problem. Therapy radionuclides may require different facilities than radionuclides used for diagnostic procedures, to ensure the safe preparation and administration of the radiopharmaceutical.

In both diagnostic and therapeutic nuclear medicine, the patient becomes a source of radiation not only for him/herself but also for staff, caregivers, and the

S. Mattsson (✉)
Medical Radiation Physics, Department of Clinical Sciences Malmö, Lund University, Skåne University Hospital, 205 02 Malmö, Sweden
e-mail: soren.mattsson@med.lu.se

S. Mattsson and C. Hoeschen (eds.), *Radiation Protection in Nuclear Medicine*,
DOI 10.1007/978-3-642-31167-3_1,

general public and remains so until the radioactive material has decayed or is excreted from the body. Personnel involved in nuclear medicine must have good knowledge of radiation protection. This is vital for patient safety as well as for the staff's own security.

For patients, radiation protection is ensured (1) by performing only those tests and treatments that are necessary (*justification*) and (2) by *optimization*, using the best radiopharmaceuticals, optimally adjusted and calibrated equipment to provide the best test results or treatment outcomes, using standard tests, procedures, and administrative controls, and having knowledgeable and trained personnel. The overriding principle is that any test or treatment should offer the maximum benefit to the patient and limit the radiation exposure.

When considering the justification for a medical exposure, the benefit is weighed against the detriment, including radiation effects. For diagnostic procedures, the potential detriment is the risk of inducing cancer. This risk is greater in children and decreases with age. For adults, the overall lifetime risk of fatal cancer is estimated to be 5% per Sv [1].

For an effective dose of 20 mSv, the nominal risk is about 1 in 1,200 for adults aged 30–60 years at the time of exposure. For adults aged 70 or more, the risk falls to <1 in 3,000. However, for children up to 10 years old, the risk is about 1 in 450 [1]. Most diagnostic procedures expose the patient to considerably less than 20 mSv.

Once clinically justified, each diagnostic examination should be conducted so that the dose to the patient is the lowest necessary to achieve the clinical aim. The quality of the images and the complexity of the examination should be sufficient for the intended purpose of the procedure. Since patients may have direct benefits from the exposures, it is not appropriate to impose limits on the doses received from justified examinations.

The optimization process necessarily requires a balance between administered activity (and thus patient radiation dose) and image quality. The activity administered should be sufficient to produce acceptable image quality for the diagnostic information being sought. It is important to plan the examination, including the requirement for image quality, to fit the clinical problem. This ensures that the investigation has the best opportunity to address the diagnostic question at hand. The size and age of the patient, and the time for which the patient can comfortably remain still for the study, will influence the activity required to be administered. There are wide variations in the activity administered to patients of standard body size, suggesting that there may be scope for optimization. The implementation of *diagnostic reference* levels is a practical tool to aid in dose optimization. Repeating examinations due to poor quality of the radiopharmaceutical, incorrect administration of the radiopharmaceutical, and technical problems with the imaging equipment should be minimized. Repeated procedures may be necessary if the image does not provide the clinical information required. A comprehensive quality assurance program, which includes radiopharmacy and equipment quality control, is important to obtain optimal diagnostic information from the procedures. When radioiodinated compounds are to be administered for conditions other than thyroid disease, the use of a thyroid blocking agent should be considered for the patient in order to reduce the radiation dose to the thyroid.

For the staff, the main sources of radiation exposure are the handling of radioactive material during its compounding and administration to patients, the need to position the patients for imaging, the attending of patients who had radioactive compounds administered to them, and the operation of equipment used. Training maintains a crucial role in radiation protection. This book is a documentation based on a number of lectures at a training course in "Radiation protection in nuclear medicine" and is the second part in a series of three books about nuclear medicine physics, technology, and radiation protection. The first book in this series "Radiation physics for nuclear medicine" [2] contains a number of articles which are also relevant for radiation protection and safety: articles on radiation detectors, biokinetic models, voxel phantoms, etc. These are not repeated in the present volume. The third volume in this series will deal about "Imaging in nuclear medicine."

References

1. International Commission on Radiological Protection (2007). The 2007 Recommendations of the International Commission on Radiological Protection. ICRP Publication 103. Ann ICRP 37 (2–4)
2. Cantone MC, Hoeschen C (eds) (2011) Radiation physics for nuclear medicine. Springer, Heidelberg, pp 1–285

Part I
Measurements

Chapter 2
Dose Quantities and Units for Radiation Protection

Sören Mattsson and Marcus Söderberg

2.1 Introduction

In all fields where there is a need for quantitative measurements, it is necessary to have understandable and precise quantities and units. Practically all countries use the SI system (from French: Le Système International dÚnités). In the field of radiation dosimetry and radiation protection, two other international organisations are active in relation to quantities and units: The International Commission on Radiation Units and Measurements (ICRU), which is mainly working with the physical aspects of dosimetry, and the International Commission on Radiological Protection (ICRP), which mainly works with assessments and quantification of the biological effects of radiation and provides recommendations and guidance on all aspects of radiation protection against ionising radiation.

The goal of the current system of quantities and units is to assess the biological effects resulting from external and internal exposure to ionising radiation in terms of stochastic (cancer induction, genetic effects) as well as deterministic effects (tissue effects) in order to have sufficient mechanisms to control these effects. There are excellent summaries of the evolution of dose quantities and units [1, 2] as well as of the more recent and current situation [3–6].

The present structure of radiation protection quantities and units is complicated and difficult to be readily used in practice, where they may cause some confusion among radiation workers and even among those who are responsible for the regulatory control of occupational radiation exposure at the workplace [7].

S. Mattsson (✉) • M. Söderberg
Medical Radiation Physics, Department of Clinical Sciences Malmö, Lund University, Skåne University Hospital, 205 02 Malmö, Sweden
e-mail: soren.mattsson@med.lu.se

S. Mattsson and C. Hoeschen (eds.), *Radiation Protection in Nuclear Medicine*,
DOI 10.1007/978-3-642-31167-3_2,

The aim of this educational book chapter is to describe dose quantities used in radiation protection, their units and the relations between them. There will also be a discussion on how the currently used quantities and our way to use them could be improved.

2.2 The Basic Dosimetric Quantity

2.2.1 *Absorbed Dose*

Various dose quantities have been designed by ICRP and ICRU to meet the need to protect human beings (protection quantities) and operational dose quantities which are designed for use in radiation measurements of external irradiation. All dose quantities are based on the fundamental definition of absorbed dose in a point [8] as the quotient of $\mathrm{d}\bar{\varepsilon}$ by $\mathrm{d}m$, where $\mathrm{d}\bar{\varepsilon}$ is the mean energy imparted to matter in an infinitesimal volume $\mathrm{d}V$ at a point of interest in a material of density ρ during a certain period of time by ionising radiation and $\mathrm{d}m$ is the mass in $\mathrm{d}V$. The absorbed dose is defined as

$$D = \frac{\mathrm{d}\bar{\varepsilon}}{\mathrm{d}m}. \tag{2.1}$$

In other words, the nonstochastic quantity absorbed dose is defined as the statistical average of the energy imparted per unit mass at a point. In spite D is a point quantity, it should be recognised that the physical process does not allow $\mathrm{d}m$ to approach zero in the mathematical sense [8]. The unit of absorbed dose is the gray (Gy), and 1 Gy is equal to 1 J/kg.

To illustrate the specific nature of energy absorption when it relates to ionising radiation, it may be of interest to realise that an energy absorption of 280 J in a 70-kg person (which is equivalent to the energy in a sip of hot coffee or tea) gives a mean whole-body absorbed dose of 4 Gy (which is a lethal absorbed dose from ionising radiation).

Absorbed dose can be measured absolutely or relatively using advanced equipment, not at all suitable for daily radiation protection work. In all fields of radiation protection, there is an interest to estimate the risk to the individual or to a group of individuals of the exposure which he/she or they have undergone. Together with the basic quantity absorbed dose, there are two types of quantities defined for specific use in radiological protection: *protection quantities* (defined by the ICRP and used for assessing the exposure limits) and *operational quantities* (defined by the ICRU and intended to provide a reasonable estimate for the protection quantities). How these quantities are related and how they are related to absorbed dose are schematically illustrated in Fig. 2.1 [9]. Information in ICRU Publication 57 [10] facilitates the conversion between operational and protection quantities.

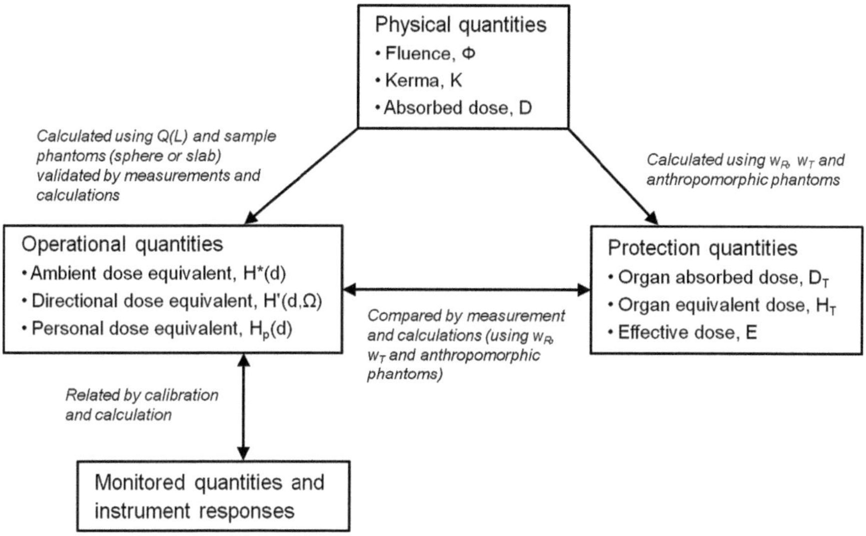

Fig. 2.1 Relationship between physical protection and operational quantities, [9]

2.3 Protection Quantities

A point quantity is not very useful for radiation protection. The average absorbed dose in a given tissue or organ is supposed to be a better indicator of the probability for radiation effects, to assess radiation exposures to humans and other organisms in a quantitative way and to describe dose/response relationships for radiation effects, the basis for risk estimation in radiological protection. This concept is based on the linear dose–effect relationship and the additivity of doses for risk assessment as an appropriate approximation in the low-dose region. Otherwise, averaging of absorbed doses in organs and tissues and adding of doses over long periods would not be an acceptable procedure. Dose distributions that are highly heterogeneous (e.g. DNA precursors labelled with tritium or Auger emitters) may need special treatment.

The protection quantities are *mean absorbed dose* in tissues or organs, *equivalent dose* in tissues or organs, and *effective dose*.

2.3.1 Mean Absorbed Dose

To assess radiation exposure to humans and correlate it with the risk of exposure, mean absorbed dose in tissues or organs is used. The absorbed dose D_T, averaged over the tissue or organ T, is defined as

$$D_T = \frac{\varepsilon_T}{m_T}, \quad (2.2)$$

where ε_T is the mean total energy imparted in a tissue or organ T and m_T is the mass of that tissue or organ.

2.3.2 *Equivalent Dose: Dose Equivalent*

Radiation protection would be a very simple indeed if the deleterious effects of ionising radiation were correlated in a very simple way, ideally linearly, with absorbed dose. Unfortunately, the results of a large body of science support the conclusion that these effects are also correlated with the types of particles and their energies.

In 1973, ICRU [11, 12] defined equivalent dose *H* which is used to take into account the fact that different particle types have biological effects that are enhanced, per given absorbed dose, over those due to the standard reference radiation taken to be 200 keV photons. This quantity has the same physical dimensions as absorbed dose. The SI unit of measure is the sievert (Sv). The concept of equivalent dose is applied only to radiation exposures received by human beings. Equivalent dose is defined as the product of *Q* and *D*, where *D* is the absorbed dose and *Q* is the quality factor at that point. The dimensionless quality factor *Q* is dependent on both particle type and energy, and for any radiation field, its value is an average over all components. It is formally defined to have a value of unity for 200 keV photons. In the 1973 system, *Q* ranges from unity for photons, electrons of most energies and high-energy muons to a value as large as 20 for α-particles (i.e. ^{4}He nuclei) of a few MeV in kinetic energy. For neutrons, *Q* ranges from 2 to >10 in the 1973 system. *Q* is defined to be a function of linear energy transfer (LET). LET is the radiation energy lost per unit length of path through a material. Different types of radiation have different LET, and X-rays, gamma rays and electrons are known as low LET radiation. Higher LET is more destructive to biological material than low LET radiation at the same dose. The radiation used in nuclear medicine is typically low LET radiation. LET is approximately equivalent to the stopping power for charged particles and is conventionally expressed in units of keV μm^{-1}. All of the radiation ultimately manifests itself through charged particles, so LET is a good measure of localised radiation damage to materials not limited to biological structures. For the common situation where a spectrum of energies and a mixture of particle types are present, the value of *Q* for the complete radiation field is an average over the spectrum of LET present weighted by the absorbed dose as a function of LET, *D* (LET).

ICRP uses radiation weighting factors w_R to connect absorbed dose to the protection quantity dose equivalent. The dose equivalent in tissues or organs is defined as

Table 2.1 Radiation weighting factors [6]

Radiation	w_R
Photons, electrons and muons of all energies	1
Neutrons	See Fig. 2.2
Protons >2 MeV (except recoil protons)	2
Alpha particles and heavy ions	20

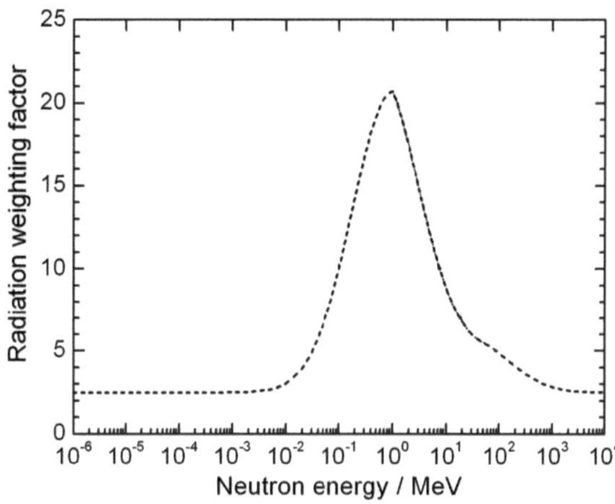

Fig. 2.2 Radiation weighting factors, w_R, for external neutron exposure for neutrons of various energies [6]

$$H_T = \sum_R w_R D_{T,R}. \tag{2.3}$$

The values of the radiation weighting factors w_R are given in Table 2.1. In general, the values of w_R in the 1990 system [9] are larger than those of Q used in the 1973 system. The latest guidance found in ICRP Publication 103 [3] has reduced the size of this increase in w_R for neutrons in some energy intervals (Fig. 2.2).

2.3.3 *Effective Dose*

The quantity effective dose (originally named effective dose equivalent) was introduced to solve conceptual and practical problems (in particular for internal irradiations) with until then used limitation concept based on "critical organ" and "maximum permissible dose". ICRP defines effective dose as [6]:

Table 2.2 Organ/tissue weighting factors [6]

Organ/tissue	w_T
Bone marrow, colon, lung, stomach, breast, remainder[a]	0.12
Gonads	0.08
Bladder, liver, oesophagus, thyroid	0.04
Bone surface, skin, brain, salivary glands	0.01

[a]Mean for adrenals, extrathoracic airways, gallbladder, heart, kidneys, lymph nodes, skeletal muscle, oral mucosa, pancreas, SI, spleen, thymus, prostate/uterus–cervix

$$E = \sum_{\mathrm{T}} w_{\mathrm{T}} H_{\mathrm{T}}. \tag{2.4}$$

The values of the tissue weighting factors w_T are given in Table 2.2. The concept of "critical organs" could be abandoned as enough knowledge became available to calculate a weighted whole-body dose. The weighting procedure was first described in ICRP Publication 26 [13], but the new quantity was not presented. In a statement from the 1978 Stockholm Meeting of the ICRP [14], the effective dose equivalent was introduced following a proposal by Wolfgang Jacobi. There are a number of assumptions, simplifications and approximations included in the definition of effective dose. It assumes validity of the LNT (linear non-threshold) model in the low-dose range and validity of temporal additivity of dose (committed dose) in the low-dose range.

In 1991, ICRP [5] replaced effective dose equivalent H_E by the effective dose, E. There was no conceptual change compared to $H_{E'}$, but the effective (organ) quality factors were replaced by radiation weighting factors; the number of tissues and organs taken into account was increased, and the values for some tissue weighting factors were modified.

The organ doses that are needed for the calculation of effective dose are calculated for reference male and female persons using a family of *reference phantoms* of which the adult male and the adult female phantom is already published [15] and phantoms for children and newborn are under way.

Using the reference phantoms, *dose conversion coefficients* for external and internal exposure are calculated for *reference conditions*:

- Standard irradiation geometries for external radiations [16]
- Standard (ICRP) biokinetic models for internal emitters, e.g. [17].

Effective dose is *not* based on data from any one individual person and does not provide an individual-specific dose but rather that for a *reference person* under a given exposure situation. Effective dose is therefore, and because of the underlying approximations and simplifications, not suitable for risk assessments for individuals. It is however of practical value for comparing the relative doses related to stochastic effects from different diagnostic examinations, the use of similar technologies and procedures in different hospitals and countries and the use of different technologies for the same medical examination, provided that the representative patients or patient populations for which the effective doses are derived are similar with regard to age and gender.

2.4 Operational Quantities

The human body-related protection quantities, equivalent dose in an organ/tissue and effective dose, are not measurable. To overcome these practical difficulties for external photon irradiation, ICRU [18–20] has introduced and defined a set of *operational quantities*, which can be measured and which are intended to provide a reasonable estimate for the protection quantities. These quantities aim to provide a conservative estimate for the value of the protection quantity avoiding both underestimation and too much overestimation. The operational quantities are based on point doses determined at defined locations in defined phantoms.

One such phantom is the ICRU-sphere [21]. It is a sphere of 30 cm diameter with a density of 1 g/cm^3 and a mass composition of 76.2% oxygen, 11.1% carbon, 10.1% hydrogen and 2.6% nitrogen.

For practical calibration work, the ICRU-sphere can be replaced by a square block with the same composition and with the dimensions 30 cm $\times$ 30 cm $\times$ 15 cm.

A single depth in the ICRU-sphere (or square block) has been recommended at which a practical approximation of the effective dose to an adult can be obtained.

For conceptual simplicity and for practicality of measurement, the operational quantities are defined (ICRU Report 39) as point functions [18], i.e. their values at a specified point depend only on the radiation field at this point. Nevertheless, they are related to an extended, remotely anthropomorphic phantom—the ICRU-sphere. To resolve this apparent contradiction, the somewhat artificial concept of an *expanded field* is required; it is the uniform radiation field that agrees with the actual field at the specified point. The principal quantity for area monitoring is, moreover, designed to be independent of the angular distribution of the radiation field, which requires the further abstraction of an *aligned, expanded field*. This is the uniform, unidirectional field that has the same fluence distribution as the expanded field; see Fig. 2.3. Using these two auxiliary concepts, one can define a quantity for the environmental monitoring of penetrating radiation [22, 23].

2.4.1 *Ambient Dose Equivalent*

The ambient dose equivalent, $H^*(10)$, is the operational quantity for *area monitoring*. It is the dose equivalent at a point in a radiation field that would be produced by the corresponding expanded and aligned field in a 30-cm-diameter sphere of unit density tissue (ICRU-sphere) at a depth of 10 mm on the radius vector opposing the direction of the aligned field. An oriented and expanded radiation field is an idealised radiation field which is expanded and in which the radiation is additionally oriented in one direction. How radiation protection instruments are calibrated in this quantity is described below.

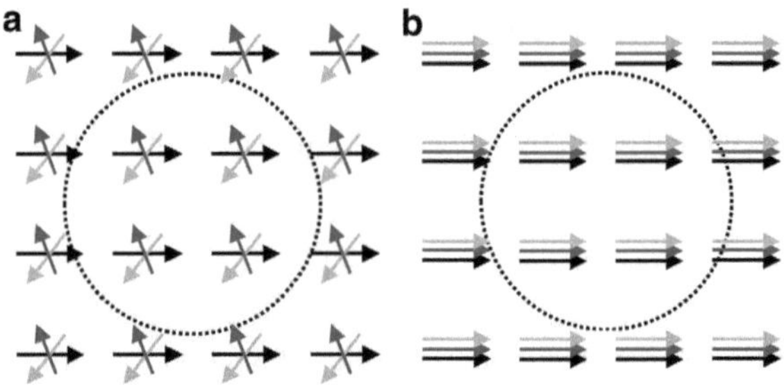

Fig. 2.3 Diagram of an expanded (**a**) and an oriented (**b**) radiation field (after [9])

2.4.2 The Directional Dose Equivalent

The directional dose equivalent, $H'(d, \Omega)$, is the operational quantity for determination of equivalent dose to skin, lens of the eye, etc., also for β-radiation and low-energy photons.

The directional dose equivalent at the point of interest in the actual radiation field is the dose equivalent which would be generated in the associated expanded radiation field at a depth of d mm on the radius of the ICRU-sphere which is oriented in the fixed direction Ω. The point is lying on a radius which has the direction Ω.

In the expanded field, the fluence and its angular and energy distribution have the same values throughout the volume of interest as the actual field at the point of reference. In the aligned and expanded field, the fluence and its energy distribution are the same as in the expanded field, but the fluence is unidirectional. If the dose equivalent in skin is to be determined, the depth 0.07 mm shall be used. If the dose to the lens is to be estimated, a depth of 3 mm might be better, while 10 mm is better for deeper laying organs. The direction can be given as an angle α in relation to a reference direction.

2.4.3 Personal Dose Equivalent

The personal dose equivalent, $H_p(d)$, is the operational quantity for *individual monitoring*: the dose equivalent in soft tissue (ICRU-sphere) below a specified point on the body at an appropriate depth d.

This quantity can be used for measurements of superficial and deep organ doses, depending on the chosen value of the depth in tissue. The depth d is expressed in millimetres, and ICRU recommends that any statement of personal dose equivalent should specify this depth. For superficial organs, depths of 0.07 mm for skin and 3 mm for the lens of the eye are employed, and the personal dose equivalents for those depths are denoted by $H_p(0.07)$ and $H_p(3)$, respectively. For deep organs and the control of effective dose, a depth of 10 mm is frequently used, with the notation $H_p(10)$.

The personal dose equivalent varies from person to person and from location to location on a person, because of different scattering and attenuation. However, $H_p(d)$ can be assessed indirectly with a thin, tissue equivalent detector that is worn at the surface of the body and covered with an appropriate thickness of tissue equivalent material. ICRU recommends that dosimeters be calibrated under simplified conditions on an appropriate phantom.

2.5 Relationship Between Quantities for Radiological Protection and Monitoring Purposes

The relationship between the physical, protection and operational quantities is illustrated in Fig. 2.4. They are discussed more fully in ICRP [24] Publication 74, which provides conversion coefficients for use in radiological protection against external radiations.

There is an acceptable agreement between the operational and protection quantities for radiation fields of practical significance when the operational quantities are based on the *Q*/LET relationship given in ICRP Publication 60. In most practical situations, dosimeters provide reasonable approximations to the personal dose equivalent, $H_p(d)$, at least at the location of the dosimeter. When the exposure of the body is relatively low and uniform, it is common practice to enter the dosimeter reading, suitably calibrated, directly into the dose records as a surrogate for effective dose. However, the personal dose equivalent generally overestimates the effective dose.

For many practical situations involving relatively uniform exposure to fairly high-energy gamma radiation, the degree of overestimation is modest; for exposure to low-energy gamma or X-rays, the overestimation can be substantial. For photon energies below ~50 keV, the effective dose can be overestimated by a factor of 2, depending on the orientation of the body.

For exposure to spatially variable radiation fields or where there is partial shielding of the body or extreme variations in the distances of parts of the body from the source, the relationships between the dosimeter measurement and the effective dose are more variable and complex. The practical convention is usually that all quantitative results reported by monitoring services represent the average absorbed dose in the whole body (or the effective dose). It is further assumed that the dose from normal natural background radiation has been subtracted from the reported results. It is also assumed that medical radiation exposures have not been included in the occupational exposure. It is almost always the reading from the dosimeter, suitably modified by calibration factors, that is reported, without considering its relationship to the absorbed doses in the various organs and tissues of the body or to the effective dose. Where exposure of the body is very non-uniform (especially in medical practice) or where exposure

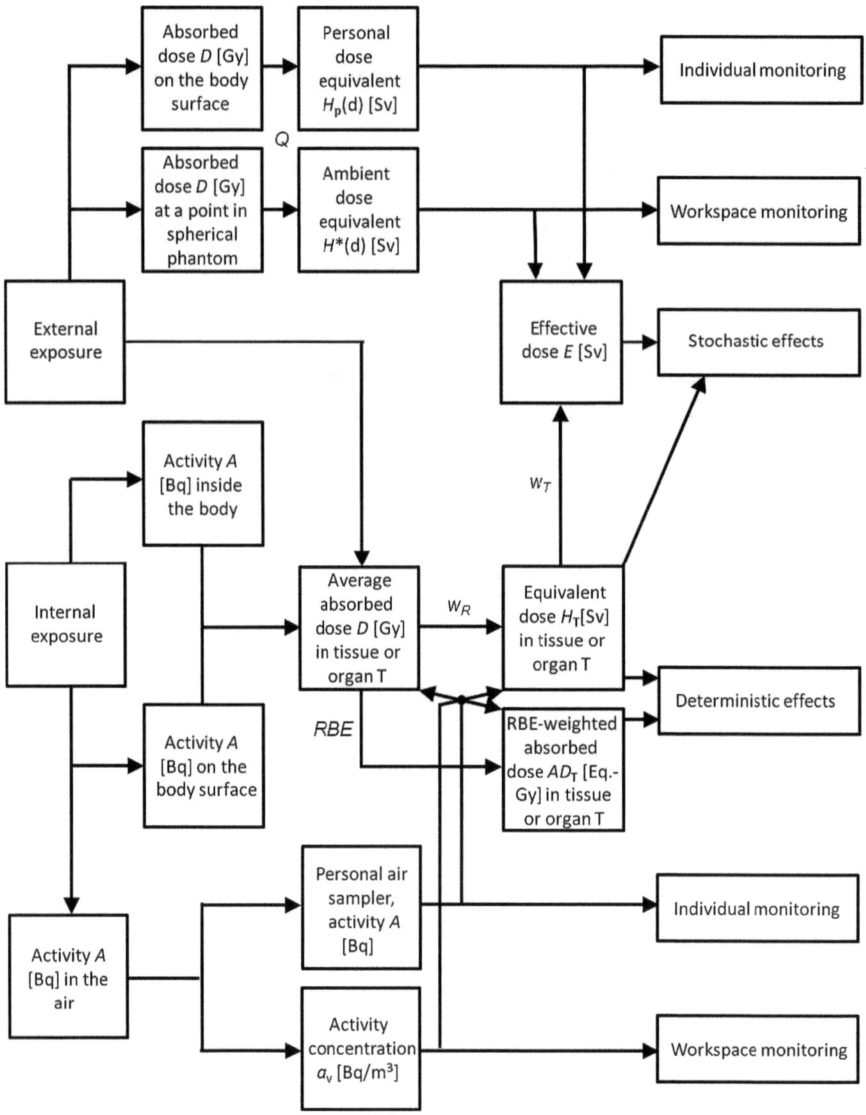

Fig. 2.4 Main quantities used in the present system of radiation protection quantifying the exposure in terms of stochastic as well as deterministic effects due to both external and internal radiation (after [9])

is mainly to low-energy radiation, the use of this convention may result in an overestimate of effective doses, which then needs appropriate qualification.

In the particular case of a unidirectional field, the direction can be specified in terms of the angle α between the radius opposing the incident field and the specified radius.

Table 2.3 The different operational dose quantities and their use (after [6])

Task	Operational dose quantities	
	Area monitoring	Individual monitoring
Estimate of effective dose	Ambient dose equivalent, $H^*(10)$	Personal dose equivalent, $H_p(10)$
Estimate of dose to the skin, hands, feet and to the lens of the eye	Directional dose equivalent, H' (0.07, Ω)	Personal dose equivalent, $H_p(0.07)$

Instruments, which are used to measure ambient dose equivalent, shall have an isotropic response. Instruments, which are used to measure directional dose equivalent and personal dose equivalent, shall have a defined directional response. Examples of detectors that can measure the ambient dose equivalent are ionisation chambers and GM tubes. Also passive detectors, like TLDs, can be used [25].

2.5.1 Calibration Methods

Most individual dosemeters and area monitors are calibrated in terms of the ICRU operational dose quantities for external exposure as part of a legal radiation protection metrology Table 2.3. There are however a number of deficiencies and limitations of the system [26].

There is a problem already in the fact that the ICRU-sphere cannot be fabricated which means that calculated conversion coefficients cannot be experimentally verified. The four-element tissue sphere is not a good phantom when calculating conversion coefficients for neutrons at lower energies. The ICRU-sphere is intended as a simple model for the human body. There are however also a need for the phantoms/models as for extremity monitoring and for the lens of the eye. In fact, there are difficulties with the definition of all three ICRU operational quantities, $H^*(10)$, $H'(0.07)$ and $H_p(d)$.

References

1. Jennings AW (2007) Evolution over the past century of quantities and units in radiation dosimetry. J Radiol Prot 27:5–16
2. Sabol J, Weng PS (1995) Introduction to radiation protection dosimetry. World Scientific, Singapore
3. Dietze G (2001) Dosimetric concepts and calibration of instruments. IRPA 10 Hiroshima 2001. http://www.irpa.net/irpa10/pdf/E03.pdf
4. Wernli C (2004) External dosimetry: operational quantities and their measurement. In: 11th International Congress of IRPA Madrid, May 2004. http://w3.tue.nl/fileadmin/sbd/Documenten/IRPA_refresher_courses/External_Dosimetry__Operational_Quantities_and_their_Measurement.pdf.

5. Dietze G, Menzel HG (2004) Dose quantities in radiation protection and their limitations. Radiat Prot Dosim 112(4):457–463
6. International Commission on Radiological Protection (2007). The 2007 recommendations of the international commission on radiological protection. ICRP Publication 103. Ann ICRP 37(2–4)
7. Sabol J, Navratil L, Rosina J (2011) Occupational exposure control: the problem of quantities in radiation protection. Radiat Prot Dosim 144(1–4):119–123
8. International Commission on Radiation Units and Measurements (2011) Fundamental quantities and units for ionizing radiation (Revised). ICRU Report 85a. J ICRU 11(1a)
9. International Atomic Energy Agency (2000) Calibration of radiation protection monitoring instruments. Safety Reports Series No. 16. IAEA, Vienna
10. International Commission on Radiation Units and Measurements (1998) Conversion coefficients for use in radiological protection against external radiation. Report 57
11. International Commission on Radiation Units and Measurements. 1971 Radiation quantities and units ICRU Report 19 (Bethesda, MD: ICRU), ICRU 1973 Dose equivalent Supplement to Report 19 (Bethesda, MD, USA: ICRU)
12. International Commission on Radiological Protection (1991) 1990 recommendations of the ICRP. ICRP Publication 60. Ann ICRP 21(1–3)
13. International Commission on Radiological Protection (1977) Recommendations of the ICRP. ICRP Publication 26. Ann ICRP 1(3)
14. International Commission on Radiological Protection (1978) Stockholm meeting ICRP 1978 Principles and general procedures for handling emergency and accidental exposures of workers
15. International Commission on Radiological Protection (2009) Adult reference computational phantoms. ICRP Publication 110. Ann ICRP 39(2)
16. International Commission on Radiological Protection (2010) Conversion coefficients for radiological protection quantities for external radiation exposures. ICRP Publication 116. Ann ICRP 40(2–5)
17. International Commission on Radiological Protection (2006) Human alimentary tract model for radiological protection. ICRP Publication 100. Ann ICRP 36(1–2)
18. International Commission on Radiation Units and Measurements (1985). Determination of dose equivalents resulting from external radiation sources, Part 1, Report 39
19. International Commission on Radiation Units and Measurements (1988) Determination of dose equivalents resulting from external radiation sources—Part 2, Report 43
20. International Commission on Radiation Units and Measurements (1992). Measurements of dose equivalents from external photon and electron radiations, Report 47
21. International Commission on Radiation Units and Measurements (1980). Radiation quantities and units. Report 33
22. Kellerer AM (1985) The new operational quantities for radiation protection. Radiat Prot Dosim 12(2):79–81
23. Sabol J (2010) Letter to the Editor. Radiat Prot Dosim 142(2-4):385–386
24. International Commission on Radiological Protection (1996) Conversion coefficients for use in radiological protection against external radiation. ICRP Publication 74. Ann ICRP 26(3–4)
25. Carlsson CA, Alm Carlsson G, Lund E, Matscheko G, Pettersson HLB (1996) An instrument for measuring ambient dose equivalent, H (10). Radiat Prot Dosim 67(1):33–39
26. Bartlett DT, Dietze G (2010) ICRU operational quantities. Editorial. Radiat Prot Dosim 139(4):475–476

Part II
Radiation Biology

Chapter 3
Radiobiology for Radiation Protection

Peter Bernhardt

3.1 Radiation Interactions and Production of Free Radicals

One of the most serious effects of radiation is that it induces changes in DNA molecules [1, 2]. Radiation can cause DNA damage either directly or indirectly; see Fig. 3.1. In the direct interaction, photons or charged particles directly interact with the DNA to create an ionization event. This ionization primarily targets hydrogen molecules, which leads to a loss of hydrogen in the DNA strand. This creates a free-radical site in the DNA. Free radicals are extremely reactive; they have high tendency for interacting with the surrounding molecules. When a free-radical site in DNA interacts with a molecule that can donate a hydrogen molecule, the DNA will return to its original, correct form. However, with oxygen present, the risk is high that an oxygen molecule will interact with the free-radical site, and this creates a single-strand break (SSB) in the DNA. Oxygen is the most potent molecule for DNA damage, but other molecules, e.g., proteins, can also interact with the free-radical site and cause DNA damage.

Water is the most abundant molecule inside cells. Radiation ionizes water molecules and thereby produces free radicals. The process is called radiolysis of water. In this way, radiation can indirectly cause DNA damage. Radiolysis of water occurs as follows: first, the water molecule is ionized; then, it is split into ionized hydrogen and a free-radical hydroxyl. Next, the ionized hydrogen can interact with a free electron, and this creates hydrogen, which is another free radical. These two free radicals (the hydroxyl and hydrogen) will diffuse into the surrounding media until they react with a molecule. Thus, several different secondary products may be created,

P. Bernhardt (✉)
Department of Radiation Physics, Institute of Clinical Science at the Sahlgrenska Academy, Department of Medical Physics and Engineering at Sahlgrenska University Hospital, 413 45 Gothenburg, Sweden
e-mail: peter.bernhardt@gu.se

S. Mattsson and C. Hoeschen (eds.), *Radiation Protection in Nuclear Medicine*,
DOI 10.1007/978-3-642-31167-3_3, © Springer-Verlag Berlin Heidelberg 2013

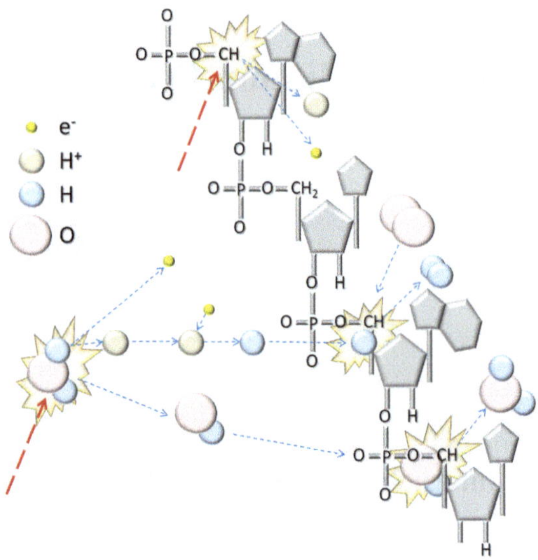

Fig. 3.1 Schematic illustration of the direct and indirect action. The *red arrow* at the top of the figure illustrates the direct interaction with the DNA molecule. The other *red arrow* illustrates the indirect interaction with the water molecule and the subsequent radiolysis

which are more or less harmful to DNA. When the induced free radicals diffuse into DNA, dehydrogenation occurs, and, as described above, a SSB may result.

The induction of SSBs depends on the oxygen level and ionization density. High oxygen pressure in tissues will increase the risk of DNA damage, either by direct or indirect action. This relationship between the oxygen level and DNA damage is not linear; however, in normal tissues, the oxygen level is nearly saturated; thus, small changes in the normal tissue oxygen pressure will cause only minor differences in DNA damage. In contrast, when the oxygen pressure is very low, small changes will cause marked changes in the number of DNA breaks. This situation occurs seldom in normal tissues, but it is frequent in tumors with poor vascularization; thus, tumors have abundant regions of poorly oxygenated cells that are less sensitive to radiation, due to the low oxygen pressure.

3.2 DNA Damage and Repair

Radiation-induced DNA damage primarily manifests as alterations in the nucleotide bases, base dimerization, base loss, hydrogen bond breaks, SSBs, and double-strand breaks (DSBs) [1, 2]. Late cellular effects are mainly caused by DSBs or more complex DNA breaks. A DSB can be caused directly by ionization events induced by a charged particle, or it can be created by two closely induced SSBs. The cell's DNA repair system can readily repair SSBs. This system comprises a panorama of enzymes. First, they recognize the damaged site; then, they unwind the DNA helix to

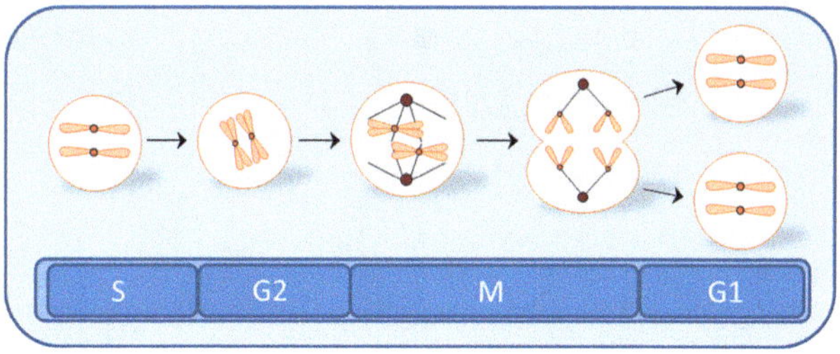

Fig. 3.2 Schematic illustration of the cell cycle ages; G1, S, G2, and mitosis (M). In the mitosis, it is demonstrated how the mitotic spindle fibers is guiding the chromosomes into different sites in the cell

provide access to the separate strands; then, they remove the damaged bases and rebuild the DNA strand, with the opposite strand for a template; finally, the DNA strand is correctly rewound to form the double-helical structure. This is a simple biological process that starts immediately after the SSB occurs, and it is finished within minutes. The repair rate depends on the specific tissue irradiated and the abundance of SSBs that are created per time unit. From a radiation protection perspective, it is worth noting that it might be more risky to receive a large dose delivered at once than to receive the same cumulative dose delivered over a long period of time. This is because, when SSBs are not repaired before a new SSB is created nearby, a DSB might form. This is important, because DSBs are much more complicated to restore; moreover, there is a risk that the DSB repair process might generate a mutation or deletion, which might later induce cancer.

There are two main repair processes for DSBs, *homologous recombination repair* (HRR), which acts after DNA duplication, and *nonhomologous end joining* (NHEJ), which acts throughout the cell cycle.

The different phases of the cell cycle are often referenced when describing cell age; see Fig. 3.2. Cell division is known as mitosis. In mitosis, the cell rearranges its components and divides into two cells. First, each chromosome pair is held together at the centromere, a molecular structure that binds sister chromosomes together while they are moved to the center of the cell. Then, chromosome pairs are separated by mitotic spindle fibers that guide one of each chromosome pair into different sides of the cell during division. After mitosis, the two daughter cells enter the cell cycle at the G1 (growth) phase in preparation for the S (DNA synthesis) phase. In the S phase, the single parent DNA is duplicated. Then, the cell enters the G2 phase to prepare for another round of mitosis. A nonproliferating cell is said to be in a G0 (resting) phase. From this phase, the cell can enter into the proliferating (G1) state when necessary. The HRR process acts when the cell has replicated its DNA. The HRR process uses the sister chromatin as a template, and DNA repair is highly accurate. The NHEJ process acts during all the other cell cycle stages and does not use the sister chromatin. For this reason, the NHEJ process is prone to making mistakes, which results in an incorrect DNA sequence.

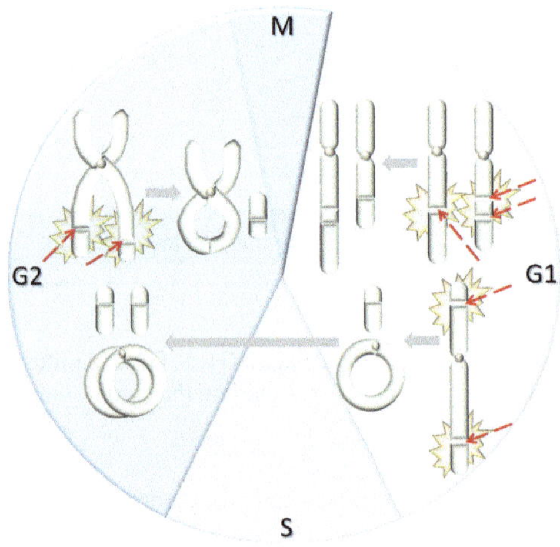

Fig. 3.3 Schematic illustration of some radiation-induced chromosome aberrations. The figure also illustrates the cell cycle ages: G1, S, G2, and mitosis (M). The small spherical structure in the chromosomes indicates the centromere. Acentric fragments are chromatids without centromere, which can be observed in the illustration, as well as ring structures, and chromosome exchanges

3.3 Chromosome Damage

Humans have a total of 46 chromosomes, which can be visualized with a light microscope. DNA damage can be observed with a light microscope only when the DNA is condensed into chromosomes, which occurs only during mitosis. However, radiation damage can sometimes be observed as a change in the chromosome structure. Deviations from the normal chromosome pattern may also indicate which stage the cell had occupied during irradiation. For example, irradiation in G1 can create dicentric chromosomes, acentric fragments, or overlapping rings [1, 2]; see Fig. 3.3.

The dicentric chromosome is generated when radiation breaks the chromosome into two fractions: one fraction contains the centromere, and the other fraction is missing the centromere. Without the intact centromere, DNA may be lost during cell division, due to an incorrect reunion of the chromosomes. A dicentric chromosome is formed when two damaged chromosomes with centromeres are joined. This new chromosome now has now two centromeres, and when it enters S phase, both strands of DNA will be duplicated to form a dicentric chromosome. Acentric fragments are created when chromosome fractions are joined together without the centromere. This will also be duplicated during S phase, and the resulting chromosome contains an acentric fragment. Overlapping rings are created from a single chromosome when radiation induces two breaks in the two arms of the chromosome. In this case, the chromosome fraction with the centromere may form a ring by joining the two arms together at the break site. This ring will be duplicated in S phase to form overlapping rings. In addition, chromosome fragments that are missing a centromere may be lost if they fail to join to another chromosome. This may represent the loss of a large amount of genetic material that might be essential for cellular control.

The loss of important genes involved in cell proliferation will cause an increase in the risk of sustaining late effects, like cancer. However, when a chromosome aberration is very large, it often leads to cell death during mitosis; the cell tries to divide but is hampered by the aberrant chromosome, and the mitotic spindle fibers are unable to guide the chromosome correctly. This unbalanced cell condition activates a protective cell signaling system that will force cell death. A well-functioning cell signaling system is crucial for assessing cellular damage and determining when to induce cellular repair and when to induce cell death. Thus, cell death can serve as an important defense mechanism for the organism to remove genetically damaged cells, and thus, it reduces the risk of cancer induction.

There are two main mechanisms for cellular death: natural, programmed cell death, called apoptosis, and forced cell death, called necrosis. In apoptosis, the signaling system informs the cell to condense its DNA content and then divide into multiple microspheres. These microspheres can enter the bloodstream and are removed by the liver. This occurs during normal cell turnover; it does not trigger inflammation processes, and the cell loss has a minor impact on the living organism. In contrast, cells that undergo mechanical or respiratory stress can die by necrosis. In necrosis, the cell swells until the cell membrane breaks, and the cellular contents are released into the surrounding tissue. This cellular debris triggers a cytokine reaction, and the organism will mount an inflammatory response.

Recent molecular biology techniques have offered insight into the molecular responses triggered by irradiation. For example, multiplex fluorescence in situ hybridization (M-FISH) allows specific labeling of the chromosomes to enhance visualization. This dramatically increased the ability to detect small chromosome aberrations. When irradiation breaks the chromosome into fragments, these fragments may rejoin correctly, they may be exchanged between chromosomes (chromosome exchanges), or they may be lost (chromosome deletions). The M-FISH technique allows dctcction of chromosome fragments that are not visible with standard light microscope techniques. In several studies, abundant exchanges between the chromosomes were observed; in some cases, chromosomes accumulated several attached chromosome fragments, and thus, severe aberrations were built into the chromosome. Accordingly, it is important to detect even small changes, deletions, and mutations in chromosomes to be able to accurately assess the damage.

3.4 Late Effects

The latest generation of techniques has enabled the detection of very small genetic aberrations and even radiation activation or inactivation of genes. Cellular studies have demonstrated that several hundreds to thousands of genes can be upregulated and downregulated after radiation exposure. The number and distributions of genes

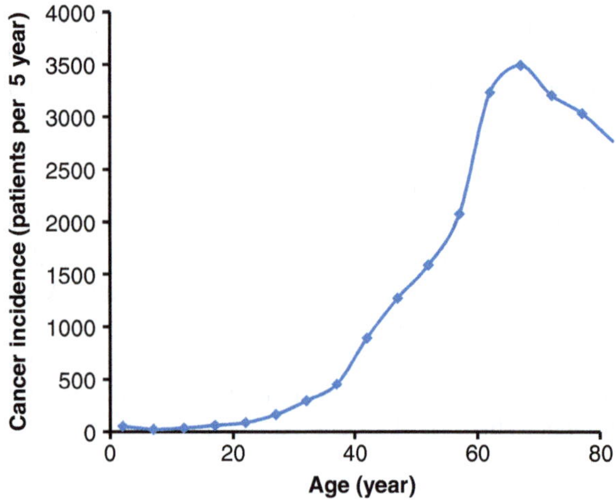

Fig. 3.4 The incidence of female cancers in the Swedish population during the year 2009 as a function of age

up- or downregulated depend on the cell type studied. An alteration in the expression of one gene can affect multiple cell signaling cascades throughout the cell. Because cell signaling is tremendously complex and diverse, alterations in many pathways may activate the signaling pathway that leads to cell death. In normal tissues, induced cell death stimulates the neighboring cells to divide. The new cells fill in the tissue gap left by the removal of dead cells. Thus, normal tissue can preserve its architecture and function. However, with each division, the cells in normal tissues experience a shortening of the telomeres at the end of the chromosomes. This limits the life of a cell, because, when all the telomeres are consumed, the cell can no longer divide. At that point, tissue gaps due to cell death can no longer be replaced by the neighboring cells. Instead, other tissue components, like collagen, will replace the missing cells. This reduces the function and plasticity of the organ, and the organism manifests signs of aging.

The aging process is most active when genetic damage stimulates successive cell divisions. Moreover, multiple genetic aberrations can induce the conversion of a normal cell into a cancerous cell. At first, the cell may retain its normal DNA with accurate repair of the damage. When this fails, the cell must detect the damage and stop cell division before it loses control and becomes a cancerous cell. Cell division can be stopped by cell death or cell senescence. Senescent cells can no longer divide. They retain function, and genetic damage can still be induced, but they no longer have the capacity to multiply. Thus, a senescent cell is not regarded as a cancer cell, and it is not fatal for the organism. When a precancerous cell is allowed to survive, it can accumulate genetic damage, and conversion into a cancerous state increases with time.

Of course, radiation is not the only factor that causes genetic damage, but it contributes to the overall risk for cancer induction. Therefore, radiation exposure should be as low as reasonably achievable whenever possible. For most cancers, the incidence increases with the age of an individual; see Fig. 3.4. This is a reflection of the fact that multiple genetic changes are required for a normal cell to be converted into a cancerous cell. The features that identify a cancerous cell include: resistance to cell death, sustained proliferation, evasion from growth suppressors, induction of angiogenesis, replicative immortality, the ability to reprogram energy metabolism, evasion from immune destruction, and the ability to invade and metastasize [3].

3.5 Radiosensitivity and Cell Survival Curves

Cell death mechanisms have been extensively studied [1, 2]. In the field of radiobiology, in vitro systems are commonly used, because radiation is easily applied and then cells can be followed in culture for successive periods of time. In these experiments, cells are cultured in conditions similar to the physiological conditions of normal tissues. Depending on cell type, cells may attach to the bottom of the culture dish, or they may float in the medium for cell growth.

In a typical cell survival experiment, the cells are seeded in culture dishes at a cell concentration sufficient to ensure a detectable number of cells at the end of the experiment. The classical quantification of cell survival involves counting cells that are capable of dividing several times to form a cluster of cells. The number of cell divisions depends on the cell line, but normally, 5–6 divisions occur within about 2 weeks. Therefore, when around 100 nonirradiated cells are seeded, after 2 weeks of culturing, spheroid clusters of cells can be noted. The number of clusters is typically less than the number of seeded cells. The plating efficiency is determined as the number of clusters divided by the number of seeded cells, times 100%. The plating efficiency of nonirradiated cells is typically between 50% and 100%.

In a typical experiment, cells are irradiated with a stepwise increase in the absorbed dose, with a maximum absorbed dose of about 10 Gy. The cells are then cultured for a few weeks, and subsequently, the survival rate is determined. The survival curve plots the logarithm of the survival fraction (because the survival fraction is frequently $<1\%$) against the mean absorbed dose. When the first cell survival experiments were performed, it was noted that the resulting survival plots often produced rounded curves. After several attempts, the mathematical model that best fit this phenomenon was the linear quadric model; see Fig. 3.5. The linear quadric model is the most common model used currently, because it is the simplest mathematical model with a reasonable fit to the data from most cell survival experiments. However, it should be noted that it is a simplified model, and caution should be taken when the model is used for extrapolation and interpolation outside the data set. Nevertheless, from a biological standpoint, the model appears to be relevant, even though it may not be exact.

The rounded cell survival curve can be divided into linear and quadric components. The linear component represents the “one target, one hit” model. With the DSB as the target, the hit is the sum of direct and indirect damage to DNA by charged particles that traverse the DNA region of interest. As mentioned above, many DSBs are repaired, and only a fraction of all DSBs lead to cell death. The production of DSBs is assumed to be correlated to the mean absorbed dose, D, and the fraction of these DSBs that lead to cell death is represented by the

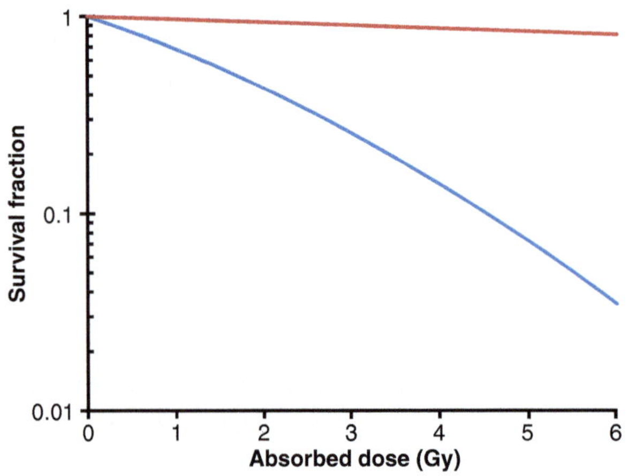

Fig. 3.5 Plot of the linear quadric (LQ) model. The *red line* is a result of the LQ model when the quadric term is missing. The survival fraction is then a pure exponential function with absorbed dose. Inclusion of the quadric term (*blue curve*) will bend the curve more rapidly with increasing absorbed doses

parameter α, which is a measure of the sensitivity of the cells to DSBs. Thus, the equation shows that the survival fraction is equal to an exponential function of absorbed dose times the radiosensitivity parameter. This component forms a straight line on the log-linear survival plot. The quadric component represents a "two hits, one target" model. Here, the target is the DSB, but the damage requires two SSBs. Again, each hit is the sum of direct and indirect damage to DNA by charged particles that traverse the DNA region of interest. In this model, a single SSB will not induce cell death; hence, another SSB must be created close to the former SSB to make a DSB. Because SSBs are rapidly repaired by the cell, the time frame between the two SSBs must be shorter than the time to repair the first SSB. Each radiation hit has a certain probability, β', for creating a SSB. This probability is correlated to the mean absorbed dose of that hit. Thus, the total probability that two hits will interact is the product of the two individual probabilities. Therefore, the survival fraction will be equal to $\beta' \times D'$ times $\beta'' \times D''$, which can be reduced to its more common form, $\beta \times D^2$. The parameter β is the radiosensitivity for inducing two SSBs that cause a lethal DSB. This component will form a curve on the log-linear survival plot. The linear quadric model is a combination of these two phenomena; thus, the survival fraction is a function of the natural logarithm with an exponent of $(-\alpha D - \beta D^2)$.

The radiosensitivity of a cell depends on the apoptotic potential, cell age, repair capacity, oxygen tension, linear energy transfer, dose rate, and absorbed dose [1, 2]. Different cell lines have exhibited widely different radiosensitivities. In particular, cells with hampered apoptotic signaling will be less sensitive to DNA damage, and the cell type will tolerate irradiation well. This type of cell displays a survival curve with a broad shoulder. In contrast, cells with highly active apoptotic signaling will be sensitive to small damages, and the cell type will readily commit apoptosis. This type of cell displays a survival curve with a small shoulder and a rapid, nearly straight decline in cell survival.

Cell radiosensitivity also varies during different cell cycle phases. To determine the radiosensitivity experimentally, cells must be synchronized in the cell cycle. Distinct G1 and G2 checkpoints allow the cell to control the cellular machinery and decide whether to continue to the next phases. The cell cycle can be stopped experimentally at the checkpoints by adding different chemicals. For example, when a chemical is added to block the cells in G1, all cells will progress through the cell cycle until they reach G1, and then, they will pause at this checkpoint. This synchronizes all the cells to the G1 phase. When the chemical is removed, the cells enter the cycle again. When the cells are irradiated at different time points, the effects on different phases can be clearly distinguished. These types of studies revealed that cells are most sensitive to radiation during mitosis, late G1, G2, and early S phase. Cells in G0 phase are often less sensitive to irradiation. This might explain why nonproliferating cells appeared to be less radiosensitive than proliferating cells.

Radiosensitivity also depends on a cell's capacity for DNA damage repair. The time required for this repair has been measured in split-dose experiments. In those studies, cells were irradiated with two doses, separated by a certain time interval. Often, cells were irradiated at low temperature to block proliferation effects, that is, cell number will not change due to cell division during the experiments. These experiments have shown that all reparable damages were repaired within a few hours; in fact, most DNA damage was repaired during the first hour. The capacity for DNA damage repair has also been tested in dose-rate experiments. There, the same cumulative dose was delivered over different times; for example, 2 Gy over 10 min is delivered four times faster than 2 Gy over 40 min. In those studies, the number of cells killed decreased as the dose rate decreased. With a slow dose rate, all reparable DNA damage could be repaired before the next DNA damage occurred. This resulted in low accumulated DNA damage, and the apoptotic signaling cascade was not triggered.

Cell radiosensitivity increases with increased oxygen pressure [4]. Experiments with different oxygen pressures have shown that non-oxygenated cells were almost 300% less radiosensitive than well-oxygenated cells; see Fig. 3.6. But, as mentioned above, in normal tissues, the oxygen level is often sufficiently high that no large differences in radiosensitivity have been observed due to variable oxygen pressure.

Linear energy transfer (LET) has a pronounced effect on the radiosensitivity of normal tissue. Low ionization, that is, low LET, is produced by gamma and electron irradiation. This radiation causes a shoulder to appear in the cell survival curve. With high-LET radiation, the shoulder seems to disappear, and the slope of the curve is considerably increased. High LET in nuclear medicine arises with alpha decay, e.g., ^{223}Ra, which is used in phase 3 trials for treating disseminated prostate cancer. The ionization around the alpha particle creates LET values around 100 keV/μm. This high-level LET will produce multiple DSBs, which increases the complexity of cellular repair. An LET value of 100 keV/μm was found to be optimal for killing cancerous cells. Higher LET values caused a reduction in the

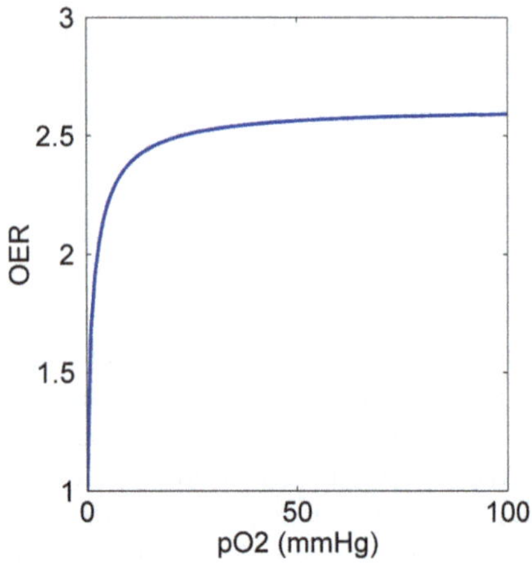

Fig. 3.6 The oxygen enhancement ratio (OER) for cell survival as a function of oxygen pressure (pO_2). The *blue line* is derived from the equation proposed by Kirkpatrick et al. In normal tissue, the pO_2 is around 40 mmHg, while tumor tissue might consist of large regions with low oxygen concentrations, making these tumor cells less radiosensitive

effect per absorbed dose. At an ionization density of 100 keV/μm, the mean distance between ionization events is around 2 nm. This is the distance between the two strands of DNA, and theoretically, maximal DNA damage will occur at this ionization level. Increasing the LET will reduce the distance between ionization events, and the DNA will be overionized, that is, further ionization will not contribute to the existing DSBs. This is consistent with the observation that high-LET radiation is insensitive to the oxygen level. In other words, the main damage to DNA is from a direct hit, and the induced free radicals create an environment that conserves damage without the need for oxygen.

Most cell survival experiments have shown that alpha emitters are five times more potent for killing cells than gamma and electron irradiation. For radiation protection purposes, the corresponding weighting factors are 20 for high- and 1 for low-LET irradiation in their potency for inducing late effects. Thus, the factor for inducing late effects is considerably higher than that for inducing acute effects. The biological reasons behind this difference between LETs are not well characterized. However, high-LET irradiation causes more complex DNA damage. Thus, we can speculate that complex DNA damage may be a more potent inducer of cancer throughout a cell's life than the DNA damage caused by low-LET irradiation.

3.6 Acute Effects

As mentioned above, the induction of cell death is a way for the organism to protect itself from potential cancerous cell growth [1, 2]. With low-dose radiation, this is probably an important step for reducing the risk of cancer induction. However, with higher absorbed doses, organs may become compromised. With abundant cell loss, organ architecture becomes rearranged, and organ function is reduced. This process progresses with accumulated damage. The first signs of organ damage are fatigue and nausea. This biological response begins at a threshold dose of 500 mGy, which is far above the achieved absorbed doses in routine nuclear medicine practice. Fatigue and nausea occur in response to the loss of cells in the crypts in the intestine, which causes intestinal bleeding. An increase in the absorbed dose to around 3 Gy will cause vomiting and diarrhea. At higher absorbed doses, damage to the intestine might elicit a gastrointestinal syndrome, which leads to death.

When the whole body is exposed to radiation, the bone marrow will set the limit for the maximal tolerated dose. Bone marrow cells are highly radiosensitive; an absorbed dose of 2 Gy or higher will dramatically reduce the number of immune system cells, e.g., granulocytes. After a high bone marrow dose, the individual will become highly sensitive to infections, but with early hospital care, bone marrow density can slowly recover, and it is nearly restored after a few months. With 5 Gy and higher exposures, the bone marrow will be completely damaged, and the only rescue option is bone marrow transplantation. At absorbed doses higher than 12 Gy, the patient will die due to cerebrovascular syndrome. The absorbed doses mentioned here are far beyond those used in nuclear medical practice. Typical absorbed doses are around 1–10 mSv/year, and additional effective doses will not impact cell number; thus, permanent organ damage is uncommon. The few cells that might be lost in protective mechanisms are effectively replaced by the growth of neighboring cells.

3.7 Summary

The human body has a built-in system for protecting against radiation exposure [1, 2]. At low exposures, the biological outcome is highly variable. The large variation in late effects stems from physical, chemical, and biological factors. A yearly effective radiation dose of around 4 mSv per year is caused by a few charged particles to traverse a single cell per year. Poisson statistics on that low number will display large variability in the number of traverses per cell; some cells will receive none at all, and others might receive ten per year. Moreover, each traverse has a variable effect in the number of ionizations that are capable of producing free radicals and direct hits to the genome. This creates a wide variation in the number of SSBs and DSBs created. The repair mechanism for the induced DNA damage also varies in sensitivity between cell types and between individuals. Apoptosis also varies in its capacity for sensing the degree of genetic damage. Taken together, these and other variations involved in transforming a normal cell into a cancerous cell make it unfeasible to accurately predict an individual's risk of developing cancer. Consequently, risk figures refer to population studies. However, increased absorbed doses will result in more complex biological damage, and cancer induction is well documented at high absorbed doses. There is evidence that at exposures to doses of ionizing radiation drawn to 100 mSv, there is an increased risk for cancer development. In some specific situations like induction of childhood cancer after irradiation of the fetus in the late stage at pregnancy, there are statistical significant increases down to around 10 mSv. According to ICRP, there is a risk of developing a lethal cancer of about 5% per Sv (0.5% per 100 mSv, 0.05% per 10 mSv, etc.) as a mean for the population, with around three times higher risk for newborn and 3–10 times lower for elderly people. In conclusion, it is our responsibility to ensure that the absorbed dose to personal and patients is as low as reasonably achievable.

References

1. Hall EJ, Giaccia A (2011) Radiobiology for the radiologist. Lippincott Williams, Philadelphia, PA. ISBN 10: 1608311937
2. Saha GB (2010) Physics and radiobiology of nuclear medicine. Springer, New York. ISBN 10: 1441921508
3. Hanahan D, Weinberg RA (2011) Hallmarks of cancer: the next generation. Cell 144(5):646–674
4. Kirkpatrick JP, Cardenas-Navia LI, Dewhirst MW (2004) Predicting the effect of temporal variations in pO_2 on tumor radiosensitivity. Int J Radiat Oncol Biol Phys 59(3):822–833

Chapter 4
Radiobiology and Radiation Dosimetry for the Lens of the Eye

Günther Dietze

4.1 Introduction

The lens of the eye is a specific part of the human body which is very sensitive to exposure by ionizing radiation, not with respect to cancer induction but mainly due to the induction of a cataract in the lens of the eye. In the past, the cataract induced by ionizing radiation has been seen as a deterministic effect (non-stochastic) with an absorbed dose threshold of 0.5–2.0 Gy for short-time exposures and 5–6 Gy for long-time exposure with low dose rate. This means that it was assumed that a cataract is not induced if the mean absorbed dose in the lens is less than about 0.5 Gy. As a consequence, the International Commission on Radiological Protection (ICRP) has not included the lens of the eye into the system of organs and tissues specified for the definition of an effective dose, E (see Table A.3.1 in ICRP Publication 103 [1]), a quantity which is mainly designed for an application at low doses where only stochastic effects, e.g. induction of cancer, and no deterministic effects are present.

Therefore, ICRP had defined a specific annual dose limit for the lens of the eye of 150 mSv for occupationally exposed persons and 15 mSv for the public and hence by values far below the dose threshold for cataract induction. Recently, however, some studies have shown [2, 3] that a cataract may be induced at even lower doses of ionizing radiation, and in its Publication 103 [1], the ICRP has mentioned that further research is seen to be necessary before deciding if annual dose limits for the lens of the eye should be lowered. During recent years, more epidemiological data became available from cataract studies on atomic bomb survivors in Japan [4, 5], from studies among Chernobyl cleanup workers [2], from occupational exposure of persons in radiology [3], and from studies on pilots and astronauts in space [6]. Based on these new data, the ICRP in a statement [7] has recently

G. Dietze (✉)
Ionising Radiation Division Physikalisch-Technische Bundesanstalt Paracelsustr. 7, 38116 Braunschweig, Germany
e-mail: guenther.dietze@t-online.de

S. Mattsson and C. Hoeschen (eds.), *Radiation Protection in Nuclear Medicine*,
DOI 10.1007/978-3-642-31167-3_4, © Springer-Verlag Berlin Heidelberg 2013

recommended to lower the limit for the lens of the eye from 150 to 20 mSv per year averaged over 5 years but not exceeding 50 mSv in a single year. This new recommendation lowers the limit by more than a factor of 7 and may have strong consequences for applications of interventional treatments in medicine where the physicians and other medical persons may achieve higher eye lens doses. Obviously, the dosimetry of the eye lens will also get more attention than before.

In the following, some information is given about the eye and its lens and the sensitivity of the lens to ionizing radiation. The modeling of the eye and its lens and the determination of doses from external exposure by electrons and photons are described. Furthermore, it is discussed which operational dose quantity is appropriate for monitoring the dose to the lens of the eye sufficiently precise for applications in radiological protection. Exposure to neutrons or heavy ions is not discussed because such exposures are restricted to very few cases, e.g. to astronauts in space or in accidental situations.

4.2 The Eye and Its Lens

It is not the aim to describe the eye and its function in detail. A schematic model of the eye is given in Fig. 4.1a. Most data of the eye geometry are from Charles and Brown [8]. The lens of the eye is small with a volume of about 0.216 cm^3 corresponding to a mass of 229 mg. The lens is positioned near the front of the eye in a depth of about 3.2 mm. The lens (see Fig. 4.1b) contains in the central part the embryonic nucleus which is surrounded by fiber cells without any inner structure (no cell nucleus, no protein structures) in order to improve the optical properties of the lens.

Epithelial cells with a cell nucleus are mainly positioned near to the front surface of the lens especially at the outer region near to the equator. While those cells are most sensitive to radiation exposure and are seen to be the target for mutations in the cell nucleus induced by the radiation, the cataract itself occurs mostly near to the back surface of the lens (cortical posterior cataract), but generally, also cataracts at other places within the lens are possible.

While for most organs and tissues ICRP recommends to use the mean absorbed dose in an organ or tissue, D_T, as the basis in radiation protection applications, the ICRP stated already in 1955 in the Supplement 6 of its first general recommendation [9]: "When the spatial distribution of radiation in the organ is very non-uniform, an average of the physical dose is not necessarily indicative of the potential damage to the organ in its relation to the normal physiological functions of the body as a whole. Therefore, in such cases it is necessary to consider a local volume within the organ in which the dose is highest. This may be called the significant volume... For the lens of the eye the significant volume is that in which the cell nuclei are located." Hence, a realistic model of the eye lens should consider this situation.

The latency time between an eye exposure and the appearance of a cataract varies from some months up to about 20 years depending on the applied dose. Low doses result in longer latency times.

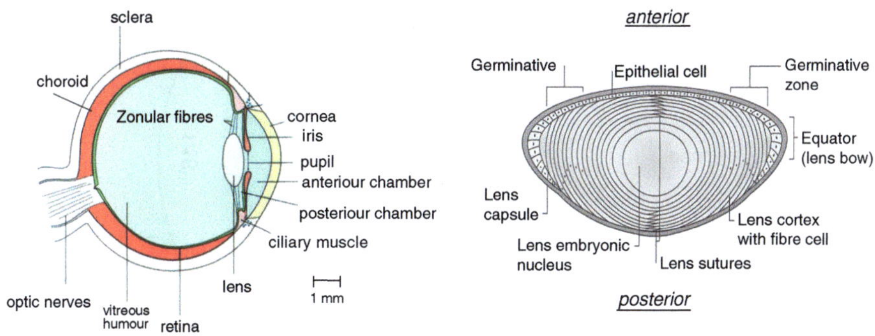

Fig. 4.1 Schematic model of the eye (**a**) and the lens of the eye (**b**)

4.3 Modeling of the Eye and Its Lens

Doses to the lens of the eye cannot be measured directly. They are generally determined by Monte Carlo calculations using appropriate models of the eye and the surrounding head. Especially for low-penetrating radiation with its small range in tissue, the modeling of the eye plays a very important role in those calculations. For example, data of conversion coefficients H_T/Φ (H_T: equivalent dose to the eye lens, Φ : fluence of incident electrons) were published in ICRP Publication 74 [10] based on data from Schultz and Zoetelief [11] which were calculated using a geometrical phantom ADAM of Kramer et al. [12] with the eye lens positioned at the surface and no material in front of the lens. Also, the reference voxel phantoms recommended by ICRP in 2007 [1] are not well suited for modeling of the eye lens. A voxel phantom simulates the human body by a large number of small volume elements (voxels) where each element is related to a specific type of tissue with a given density and atomic composition. Because of the specific selection of the voxel sizes which are too large compared to the small size of the eye lens and the determination of a mean dose value in each voxel, it is difficult to calculate eye lens doses. In addition, eyelids are not considered in both models. Hence, a more realistic geometrical model of the eye has recently been developed by Behrens et al. [13]. The main information on the geometry of the eye, its atomic composition, and densities was taken from Charles and Brown [8]. This includes also information on the most radiation-sensitive part of the lens positioned near the front surface of the lens at the outer region of the equator. Figure 4.2 shows the geometrical model in detail which also includes a simple model for the eyelids. The lids are important for dose determination in the eye, when low-penetrating radiations, e.g. electrons, are considered. For example, electrons with energies below about 1.5 MeV will not penetrate the lids. All calculations described in this chapter, however, are performed for open eyes. In the lens of the eye, a small volume is separated in order to model the position of the epithelial cells, the most sensitive region within the lens with respect to cataract induction by ionizing radiation.

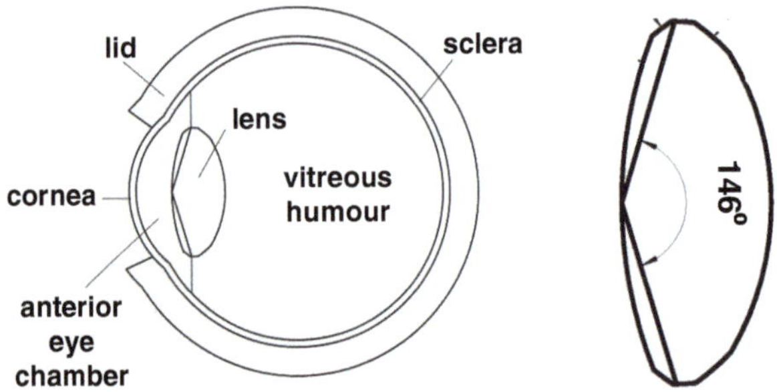

Fig. 4.2 Geometrical model of the eye (*left*) and the eye lens (*right*) from Behrens et al. [13]

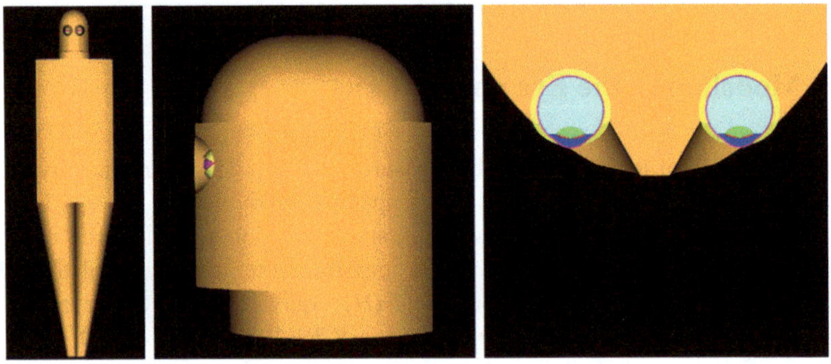

Fig. 4.3 Views of the geometry used in the Monte Carlo simulations [14]. Different colors indicate different materials. (*left*) Complete body, (*middle*) head, and eyes from the side and (*right*) head and eyes from the top with a cut at the center of the eyes

While for low-penetrating radiation, e.g. electrons, the modeling of the eye is sufficient for the calculation of doses to the eye lens, for photons and other types of penetrating radiation, the modeling of the head is also necessary in order to consider the scattering and absorption of the radiation in the surrounding tissues.

Hence, a geometrical model of the head and the body based on the geometrical ADAM and EVA phantoms of Kramer et al. [12] has been developed by Behrens et al. [14] where the medium of the head and body was simply chosen to be ICRU 4-element standard tissue [15], however, with a density of 1.11 g/cm. This simple model which is shown in Fig. 4.3 is sufficient when only scattering and absorption need to be considered and not the doses in these tissues.

4.4 Monte Carlo Calculations

There exist various Monte Carlo radiation transport codes for the calculation of mean absorbed doses in organs and tissues of the human body which are simulated by geometrically designed anthropomorphic phantoms or anthropomorphic voxel

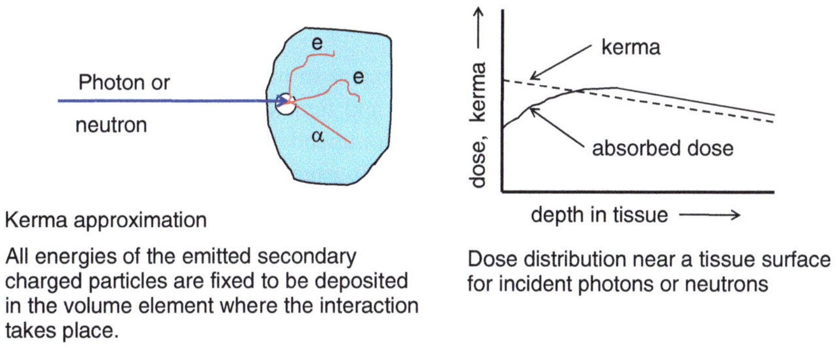

Fig. 4.4 Kerma and absorbed dose distribution near a tissue surface

phantoms. For the calculation of doses in the lens of the eye, different codes have been used.

An important point in the calculation of doses in tissue is if the codes use the "kerma approximation" or a full follow-up of the secondary charged particles in the material considered (see Fig. 4.4). Kerma approximation means that the energy transferred to the matter by the production of secondary charged particles is taken to be fully deposited in matter at the point or volume element where the interaction takes place ignoring the finite range of these particles in matter. It may be applied in cases when the incident particles are photons or other uncharged particles, e.g. neutrons. The different calculation procedures mainly result in differences near the entrance surface or near boundaries of different tissues (see Fig. 4.4). As shown below, for high-energy photons, the calculated mean absorbed dose in the eye lens strongly depends on the choice of the calculation procedure.

For dosimetric applications, often conversion coefficients are calculated using Monte Carlo codes, e.g. D_T/Φ, H_T/Φ, and E/Φ, where D_T and H_T are the mean absorbed dose and the mean equivalent dose in an organ or tissue T of the body. E is the effective dose and Φ the fluence of the radiation incident on the anthropomorphic phantom. Conversion coefficients are mainly published for simple exposure conditions assuming an external field with a constant fluence over the size of the body, e.g. frontal exposure of the body (AP), exposure from the back (AP), exposure from the left- (LLAT) or right- (RLAT) hand side, isotropic exposure (ISO), or rotational isotropic (ROT) exposure along the vertical axis of the body. For photons, the dose in the body is often related to the air kerma of the incident radiation outside of the body.

For the lens of the eye, most important data are those for AP exposure. For photons and electrons, D_T and H_T are numerically equal because the radiation weighting factor for photons and electrons of all energies is one. Data of conversion coefficients for the mean equivalent dose of the eye lens for photon and electron exposure published in ICRP Publication 74 [10] were taken from Schultz and Zoetelief [11] which were calculated using the geometrically designed ADAM phantom [12] with no tissue in front of the eye lens and the use of the kerma approximation.

Mean doses of the eye lens for photon exposure have also been calculated by Schlattl et al. [16] using the reference male and female voxel phantoms REX and REGINA defined by the ICRP in its 2007 Recommendations [1] and a full secondary charged particles follow-up. Data for electron exposure have also been obtained using the same voxel phantoms (Zankl M, private communication, 2008).

Recently, Behrens et al. [13, 14] have performed calculations for photons and electrons using the specific model for the eye described above and also a full secondary charged particle follow-up. In addition to the mean eye lens doses, the mean doses of the sensitive part of the eye lens were calculated. Results for photons and electrons are presented below.

4.5 Personal Dose Equivalent, $H_p(d)$

Mean doses of organs and tissues in the human body cannot be measured. Hence, operational dose quantities have been defined by ICRU [15] and ICRP [1] for area and individual monitoring in situations of external exposure. For individual monitoring, the personal dose equivalent, $H_p(d)$, is the quantity used in measurements with individual dosimeters worn on the body. It is defined by:

> The *personal dose equivalent*, $H_p(d)$, is the dose equivalent in ICRU (soft) tissue at an appropriate depth, d, below a specified point on the human body.
>
> The specified point is usually given by the position where the individual dosimeter is worn. For the assessment of effective dose, a depth $d = 10$ mm, and for assessing the equivalent dose to the skin, hands, and feet, a depth $d = 0.07$ mm is recommended. In cases of monitoring the dose to the eye lens, a depth $d = 3$ mm is recommended.
>
> ICRU (soft) tissue is an artificial tissue defined with a density of 1 g cm^{-3} and a mass composition of 76.2 % oxygen, 11.1 % carbon, 10.1 % hydrogen, and 2.6 % nitrogen [15].

An operational quantity for individual monitoring should provide a conservative estimate under most conditions of external irradiation. This requires that a personal dosimeter for the assessment of the dose to the eye lens must be worn at a position on the body near to the eyes, e.g. on the forehead. Calibration of individual dosimeters in terms of $H_p(d)$ is generally performed in front of standardized phantoms (see, e.g. ISO 4037-3 [17]) which simulates the backscattering of the human body. In most countries, however, the quantity $H_p(3)$ has not been introduced, and specific dosimeters for the measurements of doses to the eye lens are not yet available. In addition, there exists no international standard for the calibration of eye-lens dosimeters, and no corresponding reference phantom has yet been defined. For the calculation of conversion coefficient for $H_p(d)$ given below, therefore, the standard ISO slab and finger phantom [17] are applied. Recently, a cylinder phantom has been proposed for use in eye lens dosimetry as a cylinder much better approximates the shape of a head [18].

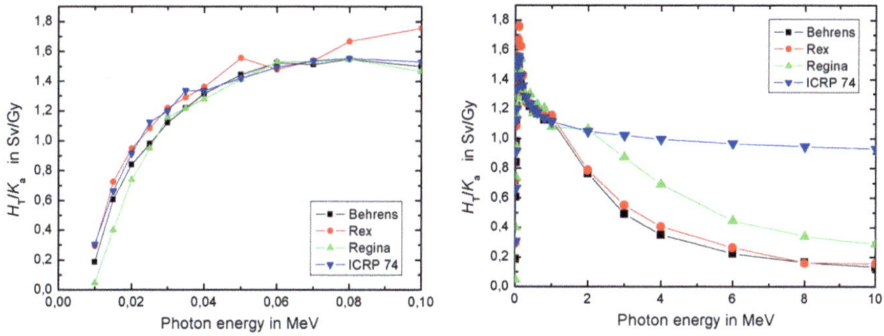

Fig. 4.5 Conversion coefficients H_T/K_a for the eye lens for frontal incidence (AP) of monoenergetic photons versus photon energy. The data are from various authors [10, 14, 16]

4.6 Photon Exposure of the Eye Lens

As mentioned above, conversion coefficients $H_{\text{eye lens}}/K_a$ for exposure by monoenergetic photons have been calculated by various authors (see Fig. 4.5). It is seen that at photon energies below about 1 MeV, the various calculations agree sufficiently well, but at higher energies, there are strong differences. The highest values are those calculated using the kerma approximation (ICRP 74 data), while the other data are absorbed dose calculations with full follow-up of secondary particles. This is a typical situation for dose calculations near the surface, especially when the calculations were performed with a phantom positioned in vacuum. Rex and Regina data differ at high energies due to the differences in the material in front of the lens for the male and female voxel phantom applied. In the following, the data calculated by Behrens et al. [14] are used for further discussion.

The question will now be discussed which personal dose quantity is best suited for measurements and for the assessment of an eye lens dose. Figure 4.6 shows the various ratios of $H_p(10)/H_{\text{eye lens}}$ and $H_p(0.07)/H_{\text{eye lens}}$ where for $H_p(0.07)$ always two different values are shown. For photon radiation, the value of $H_p(0.07)$ depends on the size of the body or phantom in front of which a dosimeter is deposited, because of the variation in the backscattering. While usually a rod phantom simulating a finger is used for calibration of dosimeters in terms of $H_p(0.07)$, for eye-lens dosimeters, a calibration in front of a larger slab phantom (30 cm × 30 cm × 15 cm) is more appropriate. The ratio of $H_p(3)/H_{\text{eye lens}}$ is not shown but would be about 1, if $H_p(3)$ would be determined on a head phantom. For a slab phantom, the ratio is higher than 1 because of higher backscattering. Obviously, the quantity $H_p(3)$ would be the first choice for use in eye-lens dosimetry. Actually, however, no type-tested dosimeters for $H_p(3)$ exist and also no international agreement on calibration procedures and phantoms. For photons with energies >30 keV, however, $H_p(0.07)$ on a slab phantom approximates $H_{\text{eye lens}}$ sufficiently precise for applications in radiation protection.

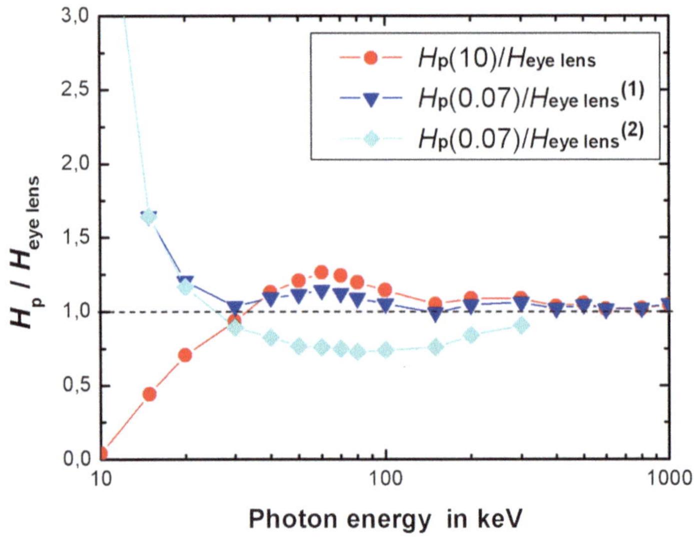

Fig. 4.6 Ratios $H_p(10)/H_{eye\ lens}$ and $H_p(0.07)/H_{eye\ lens}$ for frontal incidence (AP) of monoenergetic photons versus photon energy. $H_p(0.07)$ values obtained in the ISO slab phantom (*1*) and in the ISO rod phantom (*2*) used for calibration of dosimeters

In practice, exposures of physicians and medical staff by backscattered X-rays during interventional procedures in radiology are most important when doses to the eye lens are considered. Behrens et al. [14] have, therefore, performed some calculations by simulating a situation in interventional radiology in order to check which operational quantity can be used for monitoring the dose to the eye lens. Typically, a patient may be irradiated by x-rays from above, and medical staff may stay at the side looking to the patient.

The situation used for the calculations is schematically shown in Fig. 4.7. The body is simulated by a slab of ICRU (soft) tissue (40 cm in diameter, 15 cm thick) which is exposed by a beam of x-rays of, e.g. 100 kV (radiation qualities of RQR8 (100 kV accelerating voltage)), according to the standard IEC 61267 [19]). The beam size at the phantom surface was 20 cm in diameter. Staff is standing at the side nearby and viewing to the patient. For the calculations, the backscattered radiation under 135° to the normal axis is considered, and the spectral-photon fluence at the position of the eye is determined for a phantom-eye distance of 50 cm (more details see Behrens et al. [14]). Mean conversion coefficients for different quantities are then calculated by applying the corresponding conversion coefficients for monoenergetic photons and averaging over the photon spectrum. Figure 4.8 shows results for the quantities $H_{eye\ lens}$, $H_p(10)$, $H_p(3)$, and $H_p(0.07)$. The conversion coefficients for $H_p(d)$ are not those for a dosimeter on a head phantom but always those for the phantoms used for calibration of the individual dosimeters, hence the ISO slab phantom for $H_p(10)$ and $H_p(3)$ and for $H_p(0.07)$ both the ISO rod phantom (a) and the ISO slab phantom (b).

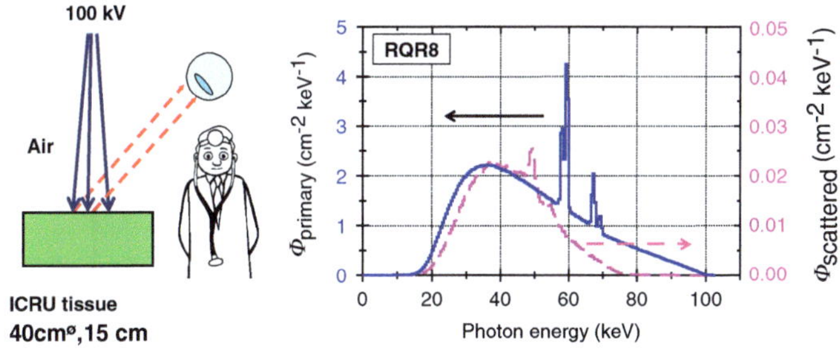

Fig. 4.7 Exposure of the eye lens by scattered x-rays. (*left side*) Schematic figure (*right side*), spectral fluence of primary, and backscattered x-rays (normalized to a total fluence of 100 cm^{-2} of the primary x-rays)

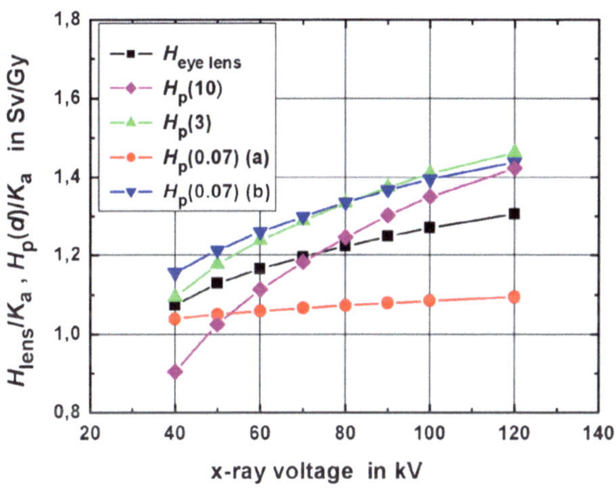

Fig. 4.8 Mean conversion coefficients $H_{\text{eye lens}}/K_a$, $H_p(10)/K_a$, $H_p(3)/K_a$, and $H_p(0.07)/K_a$ for backscattered x-rays as a function x-ray quality (data from Behrens et al. [14]). $H_p(10)$ and $H_p(3)$ data are those obtained in front of a slab phantom, $H_p(0.07)$ data are taken for a rod phantom (**a**) and for a slab phantom (**b**)

Obviously, both quantities $H_p(3)$ and $H_p(0.07)$ when calibrated on a slab phantom are appropriate for an assessment of the dose of the eye lens in x-ray radiation stray fields. At higher energies, however, $H_p(3)$ might be a better choice.

4.7 Electron Exposure of the Eye Lens

Electrons are charged particles with a relative short range in tissue. Electrons below about 0.7 MeV have a mean range of less than 3 mm in tissue (see Figs. 4.9 and 4.10) and will, therefore, not reach the lens of the eye, if electron range straggling is

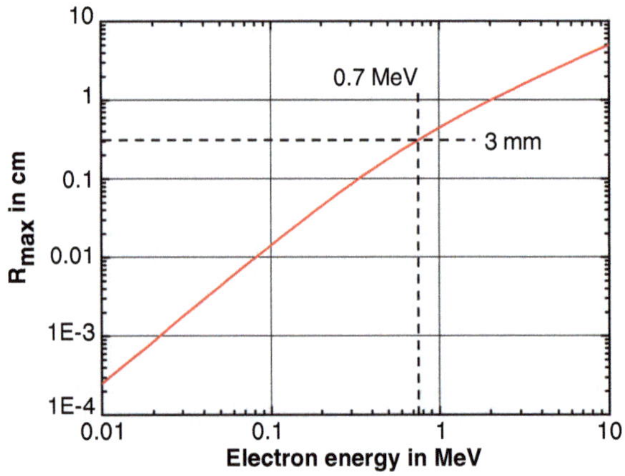

Fig 4.9 Mean range of electrons in tissue versus electron energy [20]

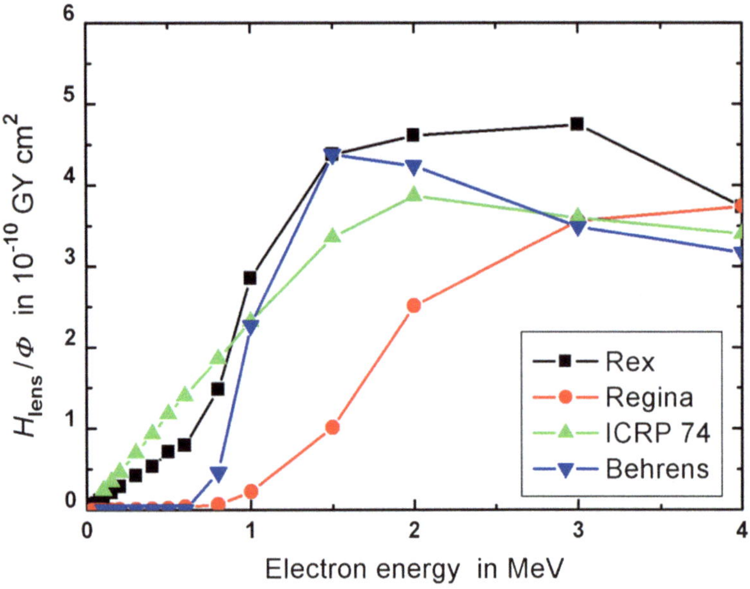

Fig. 4.10 Conversion coefficients $H_{eye\ lens}/\Phi$ for electrons versus electron energy. The data are from ICRP 74 [10], from M. Zankl using the voxel phantoms Rex and Regina (Zankl M, private communication, 2008), and from R. Behrens [13]

not considered. Dose calculations for low-energy electrons depend on the model used for the eye. Due to the different models used as has been shown in Chap. 3, dose calculations show large variations at low electron energies (see Fig. 4.10) especially at energies below 1 MeV.

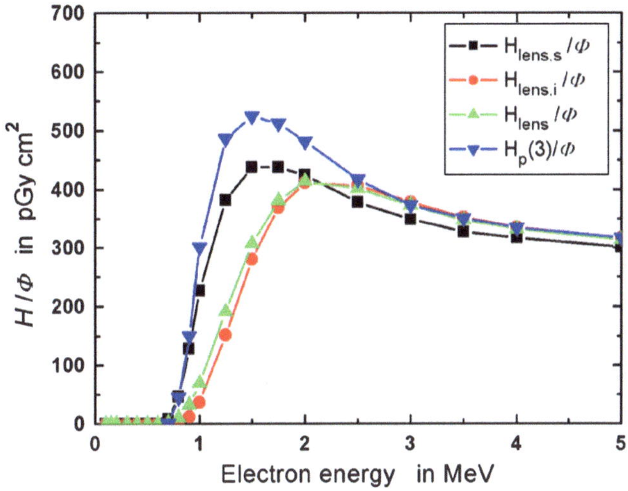

Fig. 4.11 Conversion coefficients $H_{\text{eye lens}}/\Phi$ and $H_p(3)/\Phi$ for electrons versus electron energy. The data are for the sensitive (*s*) and the insensitive (*i*) region of the lens and for the whole lens [13]. The data for $H_p(3)/\Phi$ are for the ISO slab phantom where they are equal to $H'(3,0°)/\Phi$. They are taken from ICRP 74 [10]

For further discussion, the data from Behrens et al. [13] with the detailed model of the eye are used and especially those obtained for the sensitive region of the eye lens. Figure 4.11 shows data for frontal (AP) incidence of monoenergetic electrons on the eye.

Obviously, below about 2 MeV, the data for the sensitive region of the lens agree well with the data for $H_p(3)$ especially near to the threshold between 0.7 and 1.0 MeV. This is clearly shown in Fig. 4.11, where the data for the three different operational quantities for individual monitoring are shown together with those for the eye lens. It is, therefore, recommended to use this quantity for individual monitoring in beta-radiation fields (Fig. 4.12).

All these calculations of conversion coefficients were performed with phantoms positioned in vacuum. Obviously, realistic situations are always phantoms or bodies positioned in air. This may have a strong impact on the situation with electron exposure depending on the distance between the source and the exposed phantom or body, when the energy loss of the electrons in the air cannot be ignored. Behrens et al. [21] have investigated this situation in more detail especially for realistic beta-ray sources with broad beta-energy distributions. One of the results is that for most realistic beta-ray sources, then $H_p(0.07)$ provides also a conservative assessment of the eye lens dose. For low-energy sources, however, $H_p(0.07)$ can be conservative by more than a factor of 100.

A final remark with respect to radiation protection of the eye lens is that it should always be considered that in beta-radiation fields shielding of the eye by specific spectacles can generally avoid or strongly reduce the exposure of the eye lens.

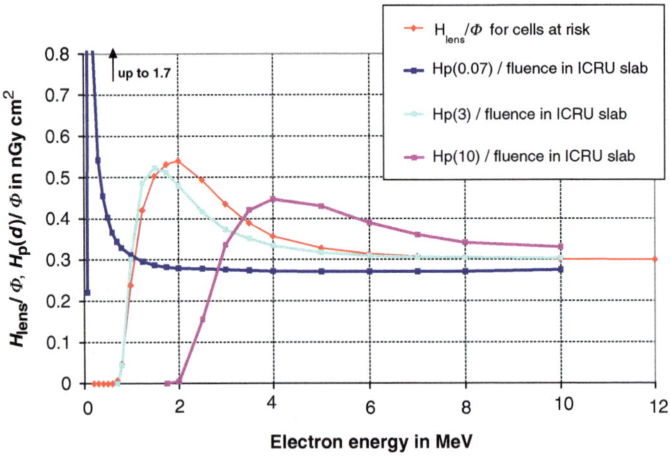

Fig. 4.12 Conversion coefficients $H_{\mathrm{eye\ lens}}/\Phi$, $H_p(10)/\Phi$, $H_p(3)/\Phi$, and $H_p(0.07)/\Phi$ for frontal incidence (AP) of electrons versus electron energy. The data for $H_p(d)/\Phi$ are for the ISO slab and were taken from ICRP 74 [10]

In photon fields, however, spectacles are not as effective as in beta-radiation fields. Sufficient distance from a radiation source is always the best way of reducing doses to the eye lens.

References

1. ICRP (2007) The 2007 recommendations of the international commission on radiological protection. ICRP Publ. 103. Ann ICRP 37(2–4):1–332
2. Worgul BV, Kundiyev YI, Sergiyenko NM et al (2007) Cataracts among chernobyl cleanup workers: implications regarding permissible eye exposures. Radiat Res 167:233–243
3. Chodick G, Bekiroglu N, Hauptmann M et al (2008) Risk of cataract after exposure to low doses of ionizing radiation: a 20-year prospective cohort study among US radiologic technologists. Am J Epidemiol 168:620–631
4. Nakashima E, Neriishi K, Minamoto A (2006) A reanalysis of atomic bomb cataract data, 2000–2002: a threshold analysis. Health Phys 90:154–160
5. Neriishi K, Nakashima E, Minamoto A, Fujiwara S, Akahoshi M, Mishima HK, Kitaoka T, Shore RE (2007) Postoperative cataract cases among atomic bomb survivors: radiation dose response and threshold. Radiat Res 168:404–408
6. Cucinotta FA, Manuel FK, Jones J, Iszard G, Murrey J, Djojonegro B, Wear M (2001) Space radiation and cataracts in astronauts. Radiat Res 156:460–466
7. ICRP (2011) Statement on tissue reactions. ICRP ref 4825-3093-1464 (Approved April 2011)
8. Charles MW, Brown N (1975) Dimensions of the human eye relevant to radiation protection. Phys Med Biol 20:202–218
9. ICRP (1955) 1954 Recommendations of the International Commission on Radiological Protection. Brit J Radiol 1955 (Suppl. 6):1–92
10. ICRP (1996) Conversion coefficients for use in radiological protection against external radiation. ICRP Publ. 74. Ann ICRP 25 (3–4)

11. Schultz FW, Zoetelief J (1996) Organ and effective doses in the male phantom ADAM exposed in AP direction to broad unidirectional beams of monoenergetic electrons. Health Phys 70:498–504
12. Kramer R, Zankl M, Williams G, Drexler G (1982) The calculation of dose from external photon exposures using reference human phantoms and Monte-Carlo methods: Part I. The male (ADAM) and female (EVA) adult mathematical phantoms. GSF Bericht S-885 (ISSN 0721-1694)
13. Behrens R, Dietze G, Zankl M (2009) Dose conversion coefficients for electron exposure of the human eye lens. Phys Med Biol 54:4069–4087 (corrigendum: Phys Med Biol 55: 3937–3945, 2010)
14. Behrens R, Dietze G (2011) Dose conversion coefficients for photon exposure of the human eye lens. Phys Med Biol 56:415–437
15. ICRU (1985) Determination of dose equivalents resulting from external radiation sources. ICRU report 39. ICRU Publications, Bethesda, MD
16. Schlattl H, Zankl M, Petoussi-Henss N (2007) Organ dose conversion coefficients for voxel models of the reference male and female from idealized photon exposures. Phys Med Biol 52:2123–2145
17. International Organization for Standardization (1999) Calibration of area and personal dosimeters and the measurement of their response as a function of energy and angle of incidence. ISO 4037-3-Part 3, Geneva, Switzerland
18. Gualdrini G, Mariotti F, Wach S, Bilski P, Denoziere M, Daures J, Bordy J-M, Ferrari P, Monteventi F, Fantuzzi E, Vanhavere F (2011) A new cylindrical phantom for eye lens dosimetry development. Radiat. Meas. 46:1231–1234
19. International Electrotechnical Commission (2005) Medical diagnostic X-ray equipment – radiation conditions for use in the determination of characteristics IEC 61267. Organization for Standardization, Geneva
20. Berger MJ et al (1998) Stopping-power and range tables for electrons, protons, and helium ions. NIST standard reference database 124, NISTIR 4999
21. Behrens R, Dietze G (2010) Monitoring the eye lens: which dose quantity is adequate? Phys Med Biol 55:4047–4062 (corrigendum: Phys. Med. Biol. 56: 511, 2011)

Part III
Radiation Dose Estimations

Chapter 5
Radiation Exposure of the Embryo/Foetus and the Newborn Child

Marie Claire Cantone

5.1 Introduction

Radiological protection distinguishes among three categories of exposures: occupational, public and medical. Medical exposures which deal with diagnostic examinations, interventional procedures and radiation therapies include patients as well as comforters, carers and volunteers in research. Medical exposures may include also exposures of the embryo/foetus of women who are pregnant or exposures of infants during breastfeeding.

The medical exposures of pregnant patients have additional ethical aspects to be considered compared to non-pregnant individual, since at least two individuals are involved in the analysis of risks and benefits. When the medical exposure is essential for the mother's life or health, there is also an indirect benefit for the foetus. Thousands of pregnant patients, and pregnant radiation workers, are exposed each year creating conditions of great anxiety for them and being, probably, responsible for a number of unnecessary terminations of pregnancies.

Most of the diagnostic procedures, in which the embryo–foetus is exposed, do not result in a measurable risk of increment over the background incidence for prenatal death, malformation, impairment of mental development or cancer induction. In general, the lifetime cancer related to the in utero exposure is assumed to be similar to the case of exposure in early childhood, and therapeutic procedures, characterised by higher doses, have the potential to result in important foetal harm.

When, for a nuclear medicine procedure, a radioactive substance, that is, the radiopharmaceutical, is introduced in a pregnant woman, that substance is distributed in her body and might reach the foetus through the placenta barrier and at a level depending on the specific substance and on the phase of the foetus development or it might be transferred into the mother's milk. The exposure of the embryo or foetus is

S. Mattsson and C. Hoeschen (eds.), *Radiation Protection in Nuclear Medicine*,
DOI 10.1007/978-3-642-31167-3_5, © Springer-Verlag Berlin Heidelberg 2013

in general the result of both the distribution of radiopharmaceuticals in the embryo or foetus and the distribution of radiopharmaceuticals in the maternal tissues.

Exposure of embryo–foetus in patients who are pregnant, as well as in workers of childbearing age, is largely considered by the different national and international bodies, interested in radiological protection of patients, workers and members of the public. For example, International Commission on Radiological Protection (ICRP) has dedicated its Publications 84, 88 and 90, respectively, to the management of pregnancy in medical radiation, to dose evaluation of embryo and foetus after intake of radionuclides by the mother and to the biological effects of prenatal irradiation [1–3]. The ICRP Publication 105 [4], within the contest of radiation protection in medicine, also discusses the need to take into consideration the termination of pregnancy after radiation exposures and the exposures of pregnant females in biomedical research, and the ICRP Publication 106 is dedicated to radiation dose to patients from radiopharmaceuticals [5], paying also attention to pregnant and breastfeeding patients.

At the European level, the Medical Exposure Directive 97/43 [6] introduces special attention to the protection of the unborn and breastfed child, exposed in the contest of medical purposes, and the Guidance Radiation Protection 100 [7], focused on unborn children and infants' exposures in relation to parental medical exposures, is addressed to prescribers, practitioners responsible for diagnosis or treatments, to medical physicists and other professional staff who are in contact with the patient.

At a national level, for example, the UK, Health Protection Agency provides guidance on the application of dose coefficients for embryo and foetus [8] and a practical guidance on how and when to prevent or to reduce unnecessary foetal exposures, when pregnant women are exposed for diagnostic purposes [9].

This chapter addresses exposures, risks and radiation protection aspects in nuclear medicine, with respect to embryo and foetus in case of pregnant patients, pregnant workers and pregnant member of a family with a family member undergoing nuclear medicine procedures. Moreover, main considerations are also included in case of breastfeeding female as patients as well as workers or member of the patient family.

5.2 Potential Effects of Radiation Exposures of Embryo and Foetus Exposures

5.2.1 The Development of Embryo and Foetus

Human prenatal development goes through three main stages: the stage of the zygote before implantation, the stage of the embryo characterised by the major organogenesis and the stage of the foetus with the predominant growth of the organs. The period encompassing the fertilisation, for forming the single-cell

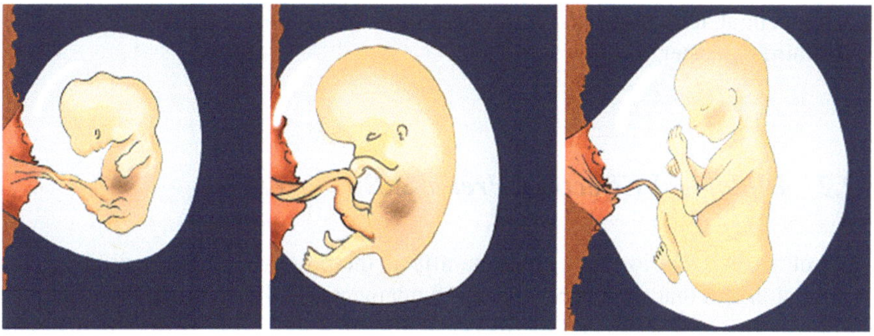

Fig. 5.1 A schematic view of the foetus development, from the *left*, at about 5 weeks, end of first trimester and beginning of third trimester

zygote, and the completion of implantation takes about 2 weeks. The implantation of the embryo into the uterine lining involves several processes giving rise to the early placental structures. The embryonic period refers to the growing organism from the second to the eighth week, taken to be 56 days after conception, and during this time, the implanted embryo develops from a tiny cell cluster to about 2–3 cm long and a weight less than about 10 g.

The embryo is connected to the placenta, which, acting as a filter and as a barrier, allows the embryo and later the foetus to absorb substances from the woman's blood and to eliminate waste from its own blood in return. Although the morphological appearance of the embryo during the first weeks of development after conception does not seem very structured, the patterns of a basic body are already established with:

1. The proliferation and differentiation of the dorsal ectodermal cells to form the neural plate and to develop into the neural tube of the nervous system.
2. The mesodermal cell layers development to form heart and circulatory system.
3. The formation of the digestive system by the endodermal cell layers.

During the second month, the embryo starts to develop a recognisable face, as well as arms, legs, fingers and toes. When the entire growing organism finally becomes clearly identifiable with human characteristics, it leaves the stage of the embryo and enters that of the foetus. The foetus period begins with the 57th day after conception, which is the beginning of the 9th week and ends at the 38th week with the birth. In this period, the rate of body growth is remarkable, especially during the 3rd and 4th months, with a development from a small growth of the embryo into a baby of about half a metre long, weighing approximately 3½ rkg. In this period, the kidneys, liver, brain and lungs are all beginning to function, the fingers and toes are separated and the external genitalia is formed. In the first weeks of this development, the male–female sex differentiation becomes apparent in the internal sex organs. Figure 5.1 is intended to give a schematic view of the foetus development at about 5 weeks, at the end of the first trimester and at the beginning of the third trimester.

5.2.2 Biological Effects of Prenatal Exposures

It is well known that ionising radiation affects the processes characterised by cell proliferation and that such processes are fundamental in the prenatal development. The sensitivity of embryo and foetus is also related to radiation effects in differentiation and cell migration as well as in the development of nervous system. Cancer risk is known to be higher after exposures of young children than of adults, and there are still questions about high radiosensitivity for prenatal exposures; in particular, there is an interest in understanding if the different development periods have different radiosensitivity and if some organs or tissues of the embryo or foetus are more sensitive to the radiation than others.

If the embryo and foetus is radiosensitive during all the prenatal period, the nature and severity of the biological effects highly depend on the development phase at the moment of exposure, and the analysis of risk for prenatal exposures is based predominantly on animal experiments and extrapolations to the characteristics of human development. On the basis of data for mice [10], the lethality and the abnormality, after prenatal exposure, at different times post-conception, show a general scheme with the highest lethality during the preimplantation period. There is a decrease in the organogenesis period and malformations after exposures during embryonic period, with no abnormalities and little lethality for exposures during foetal period.

In the following paragraphs, the effects related to radiation exposure during the preimplantation, the organogenesis and the foetogenesis will be shortly presented. The aetiology of long-term effects on brain development, the effects on neurological and mental processes and the carcinogenic risk will be also introduced.

5.2.2.1 Effects for Exposures in Preimplantation Period

Radiation effects related to exposures in preimplantation period are not observed in humans since conception is not noticed during that specific time. The related risk analysis for humans is based on data from animal experiments, mainly on mouse and rats [11]. The dominating effect in this period is the lethality effect, where doses of 0.1 Gy or less of radiation of low-linear energy transfer (LET) (see Chap. 2) can induce death mainly by chromosomal damage, both structural and numerical aberrations. In some mouse strains, the exposures to low-LET radiation and neutrons during early preimplantation period can induce malformations in the presence of genetic predisposition [12]. This kind of mechanism is still under study and evaluation [3].

5.2.2.2 Effects for Exposures in Embryonic and Foetal Periods

Short- and long-term effects are considered, since the exposures in the stages of embryonic and foetal periods determine the initially induced damage and the temporal extension of the manifestation.

For exposures during organogenesis and foetogenesis, the main categories of in utero radiation effects mainly on the basis of studies with rodents are the following: death of the embryo or foetus, malformations and growth retardation. Lethality in utero reaches a maximum after exposure to acute low-LET radiation at approximately 16 days post-conception, with a possible threshold in the dose range of 0.05–0.5 Gy. Malformations are predominantly induced during the formation of neural tube and the period of major organogenesis on about days 29–32 post-conception, with specific organ defects, in particular, skeletal malformations with a threshold in the dose range of 0.05–0.25 Gy [13]. Dose response for growth retardation is near linear for various exposure days after implantation when doses higher than 0.25 Gy are considered [14]. In the foetal period, growth retardation is higher in advanced organogenesis, with thresholds of about 0.25 Gy for the more sensitive phases of development and about 0.5 Gy for the less sensitive ones.

Radiation quality and temporal dose distribution are modifying factors of the development of biological radiation effects [3].

5.2.2.3 Effects on Brain and Mental Development

During the brain development, the radiation-induced damages are the result of (1) initial effects to the directly exposed cells (with DNA damages, cell pycnosis—a particular degeneration of the cells: the count of pyctotic cells was used in the past to study acute cell death in the prenatal rodent's brain—and apoptosis—a process of programmed cell death) and (2) secondary effects to the daughter cells and long-term response to the later cell progeny and endpoints in the mature brain.

The lowest observed dose causing persistent damages at the anatomical and structural level is in the range of 0.1–0.3 Gy of acute low-LET radiation in the period 16–25 weeks post-conception. From Japanese atomic bomb study, a severe mental retardation (SMR) after prenatal radiation exposures [15] was estimated in the most sensitive period, 8–15 weeks post-conception, with a threshold dose of about 0.3 Gy for the lower confidence bound. A radiation dose of 1 Gy could increase the risk of SMR by about 40 %. For intelligence quotient (IQ) scores, under a linear dose–response model, in the period 8–25 weeks, the reduction is evaluated about 25 IQ points/Gy [3].

5.2.2.4 Carcinogenic Risks and Hereditary Effects

The probability of such effects increases with dose, and there is no identifiable threshold below which the chance of occurrence is known to be zero. The risk of hereditary effects of ionising radiation has been estimated on basis of experiments on various animal species, because hereditary effects as a consequence of radiation exposure have not been observed in humans.

In the Oxford Study of Childhood Cancers (OSCC) [16], which is the larger study after in utero exposure, it was found that radiation increases the induction of the different kinds of tumours. The north-eastern US study [17], the second largest available study, showed a higher relative risk for leukaemia than for solid tumours, while no evidence on radiation-induced childhood cancer is found in several other studies [3].

In general, radiation has been shown to cause leukaemia and many types of cancer in both adults and children. The lifetime cancer risk following in utero exposure is assumed to have about the same risk of the exposure in early childhood, which is about three times that of the population as a whole. The estimated childhood cancer incidence is about 0.2–0.3 % for no radiation exposure above natural background. A recent analysis of epidemiological studies [1] showed a relative risk of 1.4 (a 40 % increase over the background risk) following a foetal dose of about 10 mGy. This corresponds to a probability of childhood cancer after in utero irradiation of about 0.3–0.4 %.

5.2.3 Potential Effects in the Contest of Nuclear Medicine Procedures

The considerations on effects to embryo and foetus related to prenatal irradiation, as above described and extensively presented in ICRP Publication 90 [3], are taken also as a base for judgement in the ICRP Recommendations [18] and can be summarised in few points:

1. Up to about 2 weeks after conception, the health effect of concern is the death of the embryo which is very infrequent for doses under 100 mGy.
2. During all the prenatal period, risks of noncancer health effects are not expected under 100 mGy, which is considered a sort of threshold for practical purposes.
3. The carcinogenic risk following in utero exposure is assumed to be similar to that one, following exposure in early childhood, which is about three times that of the population as a whole.

For any radiation exposure, the pregnant patient has the right to know the magnitude and type of potential radiation effects that might result from in utero exposure.

In principle, in nuclear medicine procedures, the distribution of radiopharmaceuticals in the foetus and the irradiation from the distribution of radiopharmaceuticals in the maternal tissue might contribute to the exposure of embryo and foetus. Due to low exposure to the embryo and foetus from radiopharmaceuticals in current use, as a result of diagnostic nuclear medicine procedures, these examinations present no risk of inducing foetal death, malformation, growth retardation or impairment of mental development. The main issue following in utero or childhood exposure at typical diagnostic levels of doses lower than 50 mGy is the cancer induction. When considering therapeutic nuclear medicine procedures, the potential absorbed dose to foetus is higher than in diagnostic, and risks to foetus should be estimated. From a radiation protection point of view, the justification of medical procedure involving the use of ionising radiation is required, and the optimisation of protection is also needed. Both these principles require detailed approaches and careful considerations when embryo or foetus could be involved in the medical procedure.

5.3 Radiation Protection in Case of Pregnant Patient

5.3.1 Exposure of the Foetus

Exposure to the foetus might result in general from (1) external irradiation related to radioactivity distribution in the mother's organs and tissues, (2) placental transfer of radiopharmaceuticals to foetus and (3) distribution of radiopharmaceuticals in the foetal tissues.

Most diagnostic nuclear medicine procedures make use of short-lived radionuclides (e.g. ^{99m}Tc) which therefore do not cause large foetal doses.

For radiopharmaceuticals retained by the mother, but not able to cross the placenta (e.g. radiocolloids), the exposure to the foetus is due to external sources of irradiation; thus, foetal dose is derived from the radioactivity in maternal tissues. In the case of those radiopharmaceuticals rapidly eliminated by the maternal kidneys, the major source of foetal irradiation is the mother urinary bladder. Therefore, from the foetus radiation protection point of view, the maternal hydration and frequent voiding are encouraged.

For radiopharmaceuticals (e.g. iodine isotopes) able to cross the placenta and to concentrate in a specific foetus organ, significant foetal risks could result. The physical, chemical and biological properties of the radiopharmaceuticals are the critical factors affecting the possible placental transfer.

To avoid exposure of pregnant patient during the first 2 weeks post-conception, it could be suggested to limit the non-strictly essential examinations to the first 10 days of the menstrual cycle. When a diagnostic examination is really needed for a pregnant patient, the risk to the patient of not doing that procedure is greater than the risk of potential harm to the foetus. However, in nuclear medicine, the use of some radiopharmaceuticals, typically radioiodides, can definitely increase the risk to embryo and foetus.

It is important that the female patient is interviewed by the staff to assess the likelihood of pregnancy and in any case to place in proper location inside the nuclear medicine department and health structures, illustrated advisory plates, in order to minimise the frequency of unintentional radiation exposures for the embryo or foetus. An example of such an advice is reported in Fig. 5.2.

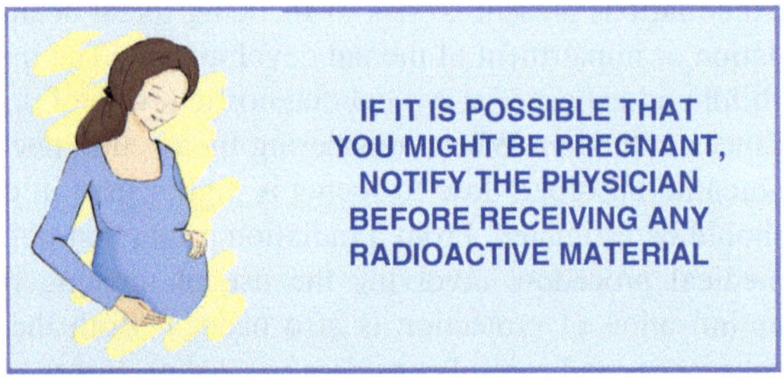

Fig. 5.2 An example for advisory notices to be clearly placed at best to be noticed inside the health-care structures

The question of termination of pregnancy in relation to radiation exposure of embryo foetus and possible related biological effects is an individual decision affected by many factors. It is in any case recognised that a dose below 100 mGy to the embryo/foetus should not be considered a reason for terminating a pregnancy. In case of doses to embryo foetus higher than 100 mGy, informed decisions should be made based upon individual circumstances. In particular, the therapy of hyperthyroidism with ^{131}I in a pregnant woman is strictly contraindicated due to possibility of external irradiation of the foetus from the radioactivity in the mother's body and moreover due to radioactive iodide crossing the placenta into the foetal circulation with subsequent uptake by the thyroid of the foetus. If a thyroid cancer with metastases is diagnosed in a pregnant woman, the treatment with ^{131}I is not compatible with the continuation of the pregnancy.

5.3.1.1 Dose Estimation to the Foetus

In general, in the first 2 or 3 months of pregnancy, the estimated absorbed dose to the embryo–foetus is substituted with the estimated dose to uterus, and moreover, for all the radiopharmaceutical, with no transfer to placenta, the dose to foetus is the same as the dose to uterus. For those radiopharmaceuticals, which might go to the foetus through the placenta, the absorbed dose to organs and tissues of the pregnant woman could be taken as a first indication of the dose to the foetal organ and tissues [5].

A detailed dose estimate to the foetus in diagnostic nuclear medicine is commonly not needed when ^{99m}Tc radiopharmaceuticals are used. When other radionuclides are used, in particular radioiodine, a more careful estimation is necessary.

The International Commission on Radiological Protection reported in its Publication 84 [1] the results of a comprehensive estimate, on the basis of a previous study by Russell et al. [19], of the foetal whole-body dose for the most common nuclear medicine examinations in early pregnancy and at term. For ^{99m}Tc radiopharmaceuticals, the reported given activities related to different procedures vary from 40 MBq for lung ventilation (aerosol) to 930 MBq for red blood cell procedure and correspondingly the reported doses range, respectively, for the two procedures and activities, from 0.1–0.3 mGy to 3.6–6.0 mGy in early pregnancy and from 0.1 to 2.5 mGy at pregnancy term. For a given patient, the absorbed dose

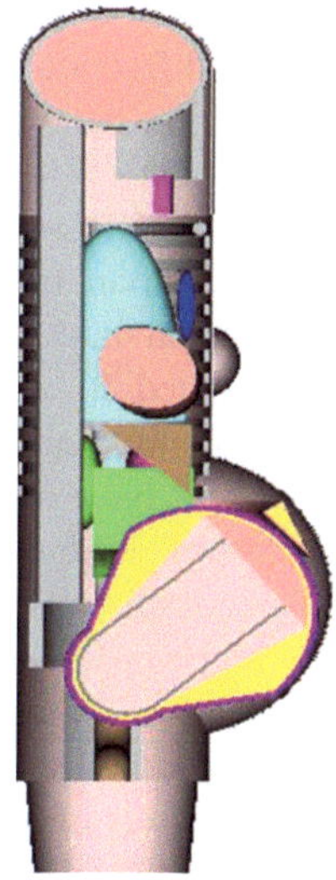

Fig. 5.3 Stylised phantom for pregnant women by Stabin et al. [22]

varies directly with the activity administered and also depends on the physical, chemical and biological properties of the radiopharmaceuticals. For example, for the ^{99m}Tc used in renal DTPA procedure, an administered activity of 750 MBq is considered, and the corresponding dose is estimated as 5.9–9.0 mGy for early and 3.5 mGy at the end of the pregnancy. Indeed, in case of DTPA, the use of a lower activity than in red blood cell procedure resulted in a higher absorbed dose. For the majority of the different radiopharmaceuticals and in each procedure, the absorbed dose to the foetus is lower than 10 mGy except for the case of using ^{131}I iodide, ^{201}Tl chloride and ^{67}Ga citrate for which the foetal dose is higher than 10 mGy [5].

In nuclear medicine, the absorbed dose is calculated, following the indications of the Medical Internal Radiation Dose (MIRD) Committee of the Society of Nuclear Medicine, by considering the energy emitted per radioactive decay, the fraction of the energy that is absorbed in the different organs which are considered as target organs with respect to organs considered as source organs, the mass of target organs, the physical decay and the biological clearance of the radioactive substances injected. The MIRD method is described by Loevinger et al. [20].

The determination of the absorbed fraction, representing the fraction of energy emitted from a source organ that is absorbed in a target organ, requires a model of the source–target configuration. Different mathematical models (phantoms) of the human body have been proposed, and standardised biokinetic models are considered. Monte Carlo methods are used to simulate radiation interactions inside the anatomic model due to the complicated geometric configuration. A set of phantoms including an adult male, five phantoms representing children of different ages [21] and four adult women, three of whom are at different stages of pregnancy, has been developed. In particular, Stabin et al. [22] modelled the changes occurring to the uterus, bladder, intestines and other organs during pregnancy and included specific models for the foetus for the calculation of absorbed fraction and specific absorbed fraction (SAF), and in Fig. 5.3, a stylized phantom for a pregnant woman is shown. The MIRD schema introduces the concept of specific absorbed fraction (SAF) defined as the absorbed fraction per unit mass of the target organ. The SAF, when combined with the data on radionuclide decay and the data on biokinetic for the considered radiopharmaceutical, allows the evaluation of absorbed doses to the embryo/foetus.

These MIRD-type phantoms have been widely used since a description is given to the human body and its organ by simple mathematical equation based on anatomical data. A more accurate series of phantoms, the voxel phantoms, is constructed in a digital 3D form, on the basis of medical images (computed tomography and magnetic resonance imaging) of real persons. Such voxel phantoms have been prepared for a wide variety of humans, from adults male and female to children and adolescents and also for pregnant women [23–25].

In consideration of the greater susceptibility, there is interest in dose estimates to the embryo–foetus from diagnostic intakes of radiopharmaceuticals by pregnant women, and research in this field is continuing. In recent years, significant developments on the techniques used in human modelling have been made, and different sets of anatomically realistic pregnant female models are under development [26, 27]: Fig. 5.4 shows the 9-month pregnant voxel model and a close-up of the foetus by Xu [28].

5.3.1.2 Foetal Thyroid Dose

Together with the average absorbed doses to the whole foetus, the evaluation of doses to individual foetal organs such as thyroid, liver and bone is desirable as higher radiation doses from activity uptake in these parts of the foetus could occur with some radiopharmaceuticals. Particular attention is required for nuclear medicine procedures using radioiodine in consideration of possible foetal thyroid uptake. In ICRP Publication 84 [1], for procedures with ^{123}I and ^{131}I by using typical administered activities of 30 and 0.55 MBq, respectively, the dose estimates for foetal whole-body range from 0.4–0.6 mGy to 0.03–0.04 mGy, respectively, for the two radiopharmaceuticals. In the same conditions, the dose estimates for foetal thyroid are considerably higher with a range from 5–10 mGy/MBq for ^{123}I to 500–1,100 mGy/MBq for ^{131}I.

A number of models [2, 29–32] are designed to calculate dose coefficient for foetal thyroid from intake of ^{131}I by the pregnant woman, and in particular, the model developed by Berkovski [31, 32] includes a physiologically realistic treatment of separate maternal and foetal iodide compartments with a bidirectional placenta transfer. Most of the models present similar tendency in the dependences of dose as a function of foetal age at time of intake [33]. For radioiodine administered to a pregnant woman, after 10–13 weeks post-conception, the foetal thyroid concentrates the iodine which has crossed the placenta barrier, and it can exceed by three- to tenfold the pregnant thyroid concentration towards the end of gestation. It can be considered that although foetal uptake increases rapidly from the 3rd month to term, the peak concentration of ^{131}I is reached at 20–24 weeks; thereafter, it decreases, because the mass of the gland increases more rapidly than

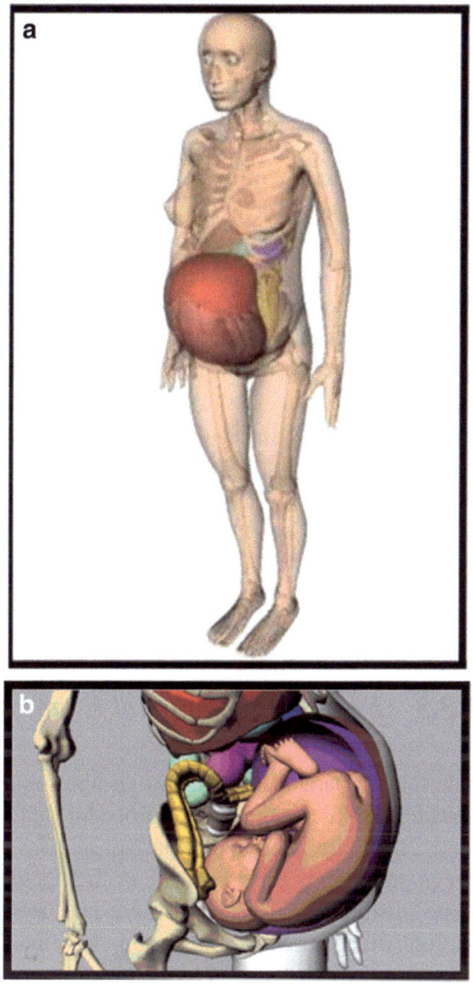

Fig. 5.4 (**a**) The 9-month pregnant model and (**b**) close-up of the foetus by George Xu [28]

the uptake does. A maximal uptake in foetal thyroid on the fifth month results in an average estimated dose of 580 mGy/MBq of ^{131}I given to the pregnant women.

5.3.2 *Therapeutic Nuclear Medicine Procedures and Pregnancy*

As a rule, pregnant patients should not be treated with radiopharmaceuticals, in particular for those rapidly crossing the placenta ^{131}I as iodide and ^{32}P as phosphate, unless the radionuclide therapy is needed in order to save mother's life. If that is the case, the absorbed dose and risk to the foetus have to be estimated by a medical physicist, a nuclear medicine physician and a radiation oncologist of the institution where the procedure is planned. The dose and risk should be communicated to the patient, and it could be a basis also for considering the termination of pregnancy. In particular, radioiodine therapy is essentially contraindicated for pregnant patients since this radiopharmaceutical crosses the placenta and the foetal thyroid tends to accumulate iodine starting from about the tenth week of pregnancy. When radioiodine therapy of thyroid carcinoma is needed, it has to be taken into consideration the need to delay the treatment after the baby's birth, but also in this case, the radioiodine excretion through breast milk has to be taken into consideration.

In relation to high dose and risk for the embryo–foetus, female patients who have the potential to be pregnant should routinely be tested for pregnancy within 72 h or less before administration of the ^{131}I treatment.

For patients who have been treated with radiopharmaceuticals, especially radioiodine therapy and radiopharmaceuticals labelled with ^{59}Fe or ^{75}Se, it is recommended not to become pregnant before the potential foetal dose, deriving from the remaining radiopharmaceutical, is reduced to <1 mGy. It is suggested to wait, after treatment, for a period of 6 or 12 months on the bases of long radionuclide physical mean time and long radiopharmaceutical biological residence time in the body and by taking into account both the reduction of radioactivity in the patient body and the case of other additional examinations needed for the treated disease.

5.3.3 Foetal Unintentional Exposure

The female patient in childbearing age should be interviewed by the staff, before starting the nuclear medicine procedure on her likelihood of pregnancy. In some cases, there are good and important reasons to administer radiopharmaceuticals to pregnant patients for diagnostic or therapeutic purposes on the basis of a properly done justification process. If there has been unintentional exposure to foetus by radiopharmaceutical administration to female patient who did not know they were pregnant, in case of ^{99m}Tc radiopharmaceuticals, a detailed dose estimate to the foetus is not normally needed, while for other radionuclides, more attention has to be given to the foetal dose and risk, in particular when considering therapeutic exposures.

Occasionally, radiopharmaceuticals have been administered for therapy to a pregnant woman, who was not aware, at that moment, she was pregnant. As the use of radiopharmaceuticals for therapy is increasing, it is important that pregnancy tests are performed. A pregnancy test by a high-detectability method is mandatory in women in fertile age before treatment, regardless of the patient's information about possibility of pregnancy [34]. Routine pregnancy testing may however give misleading results at later stage of pregnancy due to concentrations of chorionic gonadotropin being below the detection limits of these tests. Therefore, ultrasonographic examinations should be done to exclude pregnancy at the time of treatment.

A comprehensive approach following unintentional foetal exposure should include:

1. The application of dose-reduction strategies specific for the involved radiopharmaceutical like the hydration of the pregnant patient or the use of stable iodine prophylaxis.
2. The estimation of the foetal dose.
3. The information to the patient with clear explanation about the potential risks.

When foetal unintentional exposures refer to therapeutic radioiodine administration and the pregnancy of the patient became known shortly after the treatment, the hydration and frequent voiding procedure has to be encouraged to reduce the residence time of radiopharmaceutical in the pregnant patient's bladder. If the pregnancy became known after several hours, but within the first 12 h after the treatment and moreover the foetus is about 10 or more weeks old, the administration of potassium iodine as thyroid blocking agent is strongly suggested. This intervention reduces highly the result when performed with a delay exceeding the 12 h.

Consideration of terminating the pregnancy after unintentional foetal exposure is always an individual decision affected by many factors including ethical, moral and religious individual beliefs. A foetal dose below 100 mGy is not considered a reason for a pregnancy termination, and a dose higher than this level requires careful consideration about the potential foetal damage.

5.3.4 Volunteers in Biomedical Research

Biomedical researches exposing humans to ionising radiation should follow the requirements of radiation protection, and researchers have the responsibility to estimate the dose and to inform the volunteers about the related risk. Moreover, it is important to enquire on possible previous exposures of the volunteers. Ethical aspects of the participation to research, procedural issues and basis for justification are extensively discussed in ICRP Publication 62 [35]. In a number of countries, it is not explicitly prohibited to include pregnant female as subjects in biomedical research; nevertheless, it is important to consider that these kinds of research are rare and have to be discouraged, unless the condition of pregnancy is an essential part of the research.

In case of participation of pregnant volunteers in research, the radiation protection of the foetus requires detailed dose and risk estimates.

5.4 Radiation Protection in Case of Breastfeeding Patient

A number of radiopharmaceuticals are secreted in breast milk, and it is a basic and safe principle to consider that if a radiopharmaceutical is given to a breastfeeding woman, in relation to therapeutic or diagnostic nuclear medicine procedures, some

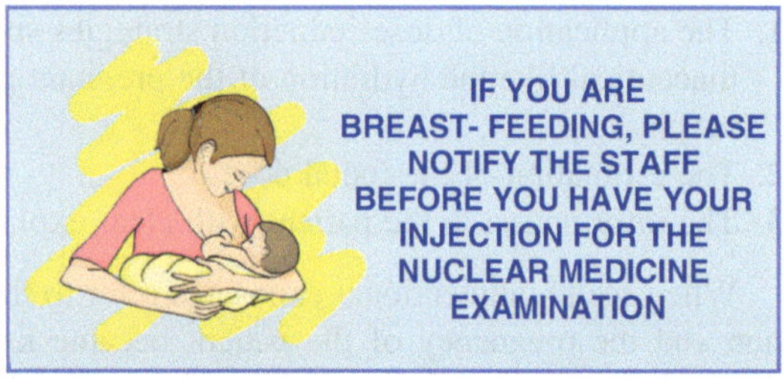

Fig. 5.5 An example for advisory notices to be clearly placed at best to be noticed inside the health-care structures

radioactivity will definitely be found in the breast milk [5, 36, 37], with consequent absorbed dose to the patient's breast and to the newborn child [38], unless data are known about the contrary for the specific radiopharmaceutical in use.

A patient who is breastfeeding should be informed and advised about the risks for the child of continuing breastfeeding, before performing any nuclear medicine procedure, and with the aim to reduce the number of potential unintentional exposure to breastfed child, it is useful to place in the department of nuclear medicine, mainly near the reception, illustrated advisory notices. An example of such a notice is reported in Fig. 5.5.

Before administering a radiopharmaceutical to a female who is breastfeeding, it should be taken into account (1) if the examination or treatment could be delayed after the ending of the breastfeeding period and (2) if the radiopharmaceutical intended to be used is the best choice, also in consideration of the aspect of secretion in breast milk. Possible substitutions to be considered in order to reduce foetal dose are:

1. ^{99m}Tc DTPA or gluconate as an alternative to pertechnetate for brain scans.
2. ^{123}I pure instead of ^{125}I or for ^{131}I.
3. ^{111}In leucocytes as alternative to ^{67} Ga for sites of infection.

When the nuclear medicine procedure is justified and then performed, possible needs to restrict breastfeeding depend on the used radiopharmaceutical and the administered activity in order to ensure that the infant will not receive an effective dose (see Chap. 2) larger than 1 mSv. In any case, where possible and useful, the mother should regularly extract and then discard the milk to maintain the milk supply.

A summary of indications about the need to stop breastfeeding for a number of radiopharmaceuticals is presented in Annex D of the ICRP Publication 106 [5]. By taking into consideration 47 of the most commonly used radiopharmaceuticals, for 12 of them (^{14} C-labelled triolein, -glycocholic acid, -urea; ^{11} C-labelled substances; ^{13} N-labelled substances; ^{15}O-labelled substances; ^{18} F-FDG; ^{51}Cr-EDTA; ^{81m}Kr-gas; ^{111}In-octreotide; ^{111}In-WBC; ^{133}Xe), the interruption is not essential, and for 15 of them, which are ^{99m}Tc-labelled, the interruption is not required, when no free pertechnetate is present in the radiopharmaceutical and only a break of 4 h, with the discard of one milk meal, should be advised. For other eight radiopharmaceuticals of the list (^{99m}Tc-labelled MAA, -microspheres (HAM), -pertechnetate, -RBC (in vivo), -WBC; ^{123}I-iodo hippurate; ^{125}I-iodo hippurate; ^{131}I-iodo hippurate), a 12-h and, for ^{201}Tl-chloride, a 48-h interruption are recommended. For the remaining 11 radiopharmaceuticals (^{123}I-BMIPP, ^{123}I-HSA, ^{123}I-IPPA, ^{123}I-MIBG, ^{123}I-NaI, ^{125}I-HAS, ^{131}I-MIBG, ^{131}I-NaI, as well as ^{22}Na-, ^{67} Ga- and ^{75}Se-labelled agents), at least 3-week interruption is recommended.

5.5 Attention to Female Workers and Members of the Public

5.5.1 Foetal Exposure for Non-patient Female in Pregnancy

In general, radiation risk is related to both external and internal irradiation, the last one normally due to the intake of radioactive substances. In the nuclear medicine field, the main source of external irradiation to other persons than patients is the radioactivity present in the patient itself, which depends on the physical and chemical characteristics of the substance and the activity administered to the patient, which is in principle higher in therapeutic treatments than in diagnostic examinations. External irradiation from a patient could involve either pregnant female workers or pregnant members of the public which are in contact with the administered patient. Intake of radioactive substance, as contamination from unsealed radioactive sources, as the radiopharmaceuticals are, could be a further exposure which could involve staff members in particular in the first period when the patient is still in hospital, and in special cases, also member of the public.

5.5.2 Issues Concerning Female Workers

Pregnancy is also an important issue for staff working with ionising radiation. It has to be remembered that for what is concerning the radiation protection of worker, the policy adopted by ICRP and other main institutions is not recommending the need to distinguish between male and female. However, if a female worker is pregnant, the protection at the working place and in the working conditions should provide to the foetus the same protection level of the public, with an additional dose to the foetus not exceeding 1 mSv during the remaining period after the pregnancy declaration by the female worker. For this reason, the working females who declared to be pregnant or breastfeeding should not be involved in radiation emergency actions.

In nuclear medicine, the staff members should pay attention in handling radiopharmaceuticals in relation to external and internal exposures, in particular for ^{131}I therapeutic doses, due to potential high dose to the operator and in radioiodination procedures due to potential inhalation risks. A female worker in pregnancy or breastfeeding should not work in areas of significant contamination risk and as an example should not work with large amounts of radioiodine.

In clinical practice, it is common that a patient is required to undergo more than one diagnostic imaging or other different procedures, and in order to prevent exposure to staff outside the nuclear medicine department, it is important that all the different procedures, where practicable, should be done before the administration of the radiopharmaceutical. Since a radioactive patient represents a radiation source, even if of low risk for other persons, proper precautions should be adopted, including the information to staff members of the other involved departments and minimising their exposures. The involved staff members of other departments should be aware about the usefulness of avoiding unnecessary proximity to the patient and of potential contamination deriving from patient body fluids and excretion, in particular in case of incontinency.

5.5.3 *Issues Concerning Pregnant Family Members, Comforters and Caregivers*

The patient administered with radiopharmaceuticals normally has family members at home, other persons taking care or giving comfort to the patient and close friends. Radiation protection for medical exposure takes into account also the protection of these individuals and in particular of children and pregnant females among them. Among the factors to be taken into account while considering the decision to keep at the hospital or to release a patient, after a therapeutic procedure, family members and home circumstances as well as the presence of infants are included.

The precaution to limit the exposure to family members and friends when the treated patient is in hospital or at home should be presented and discussed with the patient, and the possibility of being involved, at home, in taking close care of children must be considered and discussed. For the most common nuclear medicine diagnostic procedures, the total dose at 0.5 m distance from the patient ranges from 0.02 to 0.25 mGy, while at 1 m distance ranges from 0.05 to 0.10 mGy, and at these levels, there are no significant risks to pregnant members of the family [1].

On the other side, for patients treated with radioiodine, the dose, until the radioactivity complete reduction, is about 1.3 mGy for a hyperthyroid patient and 6.8 mGy for a thyroid cancer patient at 0.5 m [1]. Indeed, even more important for the family members and the caregivers is the attention to potential direct and indirect contamination deriving from patient body fluids and excretion, in particular in case of incontinency. The patients should be informed about the periods, during which restrictions should be applied to the contacts with family members and friends, and these periods are dependent on the initial external dose rate from the patient and the rate of clearance of the radiopharmaceutical from the body. As an example, by using a dose constraint of 1 mSv for children and members of the public, for a patient treated with an activity of 500 MBq of ^{131}I for thyroid cancer after thyroidectomy, a restriction of 5 days to minimise close contact with children (0–5 years old) or with pregnant women and for a patient treated with the same activity of ^{131}I for thyrotoxicosis, a restriction of 18 days is recommended for children and/or pregnant women [39].

5.6 Additional Note

This chapter is not addressing specifically the paediatric nuclear medicine procedures as extensive information is available elsewhere [40]. Indeed, as far as paediatric nuclear medicine procedures are concerned, different methods for calculating the activity to be administered have been proposed and discussed in the literature, since the fraction of the activity, to be administered to a child with respect to adult, could be based on body weight, height, surface area and age, organ size of the child and other factors [41, 42]. Moreover, in administering radiopharmaceuticals to children, it is important to take into consideration that the biodistribution and the kinetics of radiopharmaceuticals are different from adults. Recently in Europe, a dosage card, to avoid a variety of administered activities in children in different countries, was developed in the context of good practice, and North America Consensus Guidelines were achieved for a number of commonly used radiopharmaceuticals [43–45].

References

1. ICRP (2000) Pregnancy and medical radiation, ICRP Publication 84. Ann ICRP 30(1)
2. ICRP (2001) Doses to the embryo and fetus from intakes of radionuclides by the mother, ICRP Publication 88. Ann ICRP 31(1–3)
3. ICRP (2003) Biological effects after prenatal irradiation (embryo and fetus), ICRP Publication 90. Ann ICRP 33(1–2)
4. ICRP (2007) Radiological protection, ICRP Publication 105. Ann ICRP 37(6)
5. ICRP (2008) Radiation dose to patients from radiopharmaceuticals – addendum 3 to publication 53, ICRP Publication 106. Ann ICRP 38(1–2)
6. Council of the European Union (1997) Council Directive 97/43 Euratom of 30 June 1997 on health protection of individuals against the dangers of ionising radiation in relation to medical exposures. Off J Eur Commun L-180, 22–27. Luxembourg: Office for the Official Publications of the European Communities
7. Commission E (1998) Guidance for protection of unborn children and infants irradiated due to parental medical exposures. Radiation protection 100. Office for the Official Publications of the European Communities, Luxembourg
8. Guidance on the application of dose coefficients for the embryo, fetus and breastfed infants in dose assessments for members of the public. Advice from the Health Protection Agency. Documents of the HPA-Radiation, Chemical and Environmental Hazards. RCE-58, March 2000. ISBN 978-0-85951-614-3
9. Protection of pregnant patients during diagnostic medical exposures to ionising radiation. Advice from the Health Protection Agency, the Royal College of Radiologists and the College of Radiographers. Documents of the HPA-Radiation, Chemical and Environmental Hazards. RCE-9, March 2009. ISBN 978-0-85951-635-8
10. Hall EJ (1994) Radiobiology for the radiologist, 4th edn. J.B. Lippincott Co., Philadelphia, PA, USA
11. Gu Y, Kai M, Kusama T (1997) The embryonic and fetal effects in ICR mice irradiated in the various stages of the preimplantation period. Radiat Res 147:735–740
12. Streffer C, Müller W-U (1996) Malformations after radiation exposure of preimplantation stages. Int J Dev Biol 40:355–360
13. OECD (1988) The biological basis for the control of prenatal irradiation. Report by a Group of Consultants, Paris, France

14. Konermann G (1982) Consequences of prenatal radiation exposure for perinatal and postnatal development: morphological, biochemical and histochemical aspects. In: Kriegel H, Schmahl W, Kistner G, et al (eds) Developmental effects of prenatal irradiation. Gustav Fischer Verlag, Stuttgart, Germany, pp 237–250
15. Otake M, Schull WJ (1984) In utero exposure to A-bomb radiation and mental retardation. A reassessment. Br J Radiol 57:409–414
16. Gilman E, Kneale G, Knox E et al (1988) Pregnancy x-rays and childhood cancers: effects of exposure age and radiation dose. J Radiat Protect 8:3–8
17. Monson RR, McMahon B (1984) Prenatal x-ray exposure and cancer in children. In: Boice J, Fraumeni J (eds) Radiation carcinogenesis: epidemiology and biological significance. Raven, New York, USA, pp 97–105
18. ICRP (2007) Recommendations of the International Commission on Radiological Protection, ICRP Publication 103. Ann ICRP 37(2–4)
19. Russell JR, Stabin MG, Sparks RB, Watson EE (1997) Radiation absorbed dose to the embryo/fetus from radiopharmaceuticals. J Nucl Med 73:756–769
20. Loevinger R, Budinger TF, Watson EE (1991) MIRD primer for absorbed dose calculations. Society of Nuclear Medicine, New York, NY
21. Cristy M, Eckerman K (1987) Specific absorbed fractions of energy at various ages from internal photon sources. ORNL/TM-8381 V1–V7. Oak Ridge National Laboratory, Oak Ridge, TN
22. Stabin MG, Watson EE, Cristy M et al (1995) Mathematical models and specific absorbed fractions of photon energy in the nonpregnant adult female and at the end of each trimester of pregnancy. ORNL/TM-12907. Oak Ridge National Laboratory, Oak Ridge, TN
23. Zankl M, Petoussi-Henss N, Fill U, Regulla D (2003) The application of voxel phantoms to the internal dosimetry of radionuclides. Radiat Prot Dosim 105:539–548
24. Kramer R, Khoury HJ, Vieira JW, Loureiro ECM, Lima VJM, Lima FRA, Hoff G (2004) All about FAX: a female adult voxel phantom for Monte Carlo calculations in radiation protection dosimetry. Phys Med Biol 49:5203–5216
25. Petoussi-Henss N, Bolch WE, Zankl M, Sgouros G, Wessels B (2007) Patient-specific scaling of reference S-values for cross-organ radionuclides S-values: what is appropriate? Radiat Prot Dosim 127:192–196
26. Shi CY, Xu XG, Stabin MG (2008) SAF values for internal photon emitters calculated for the RPI-P pregnant-female models using Monte Carlo methods. Med Phys 35:3215–3224
27. Goo BQ, Xu XG, Shi C (2010) Specific absorbed fractions for internal electron emitters derived for a set of anatomically realistic reference pregnant female models. Radiat Prot Dosim 138(1):20–28
28. Xu XG (Rensselaer Polytechnic Institute, USA – the consortium of computational human phantoms (CCHP). http://www.virtualphantoms.org/index.html)
29. Watson EE (1992) Radiation absorbed dose to the human foetal thyroid. In: S-Stelson A, Watson EE (eds) 5th international radiopharmaceutical dosimetry symposium, CONF-910529, Oak Ridge Associated Universities, Oak Ridge, TN, USA, p 179–187
30. National Council on Radiation Protection and Measurements NCRP (1998) Radionuclide exposure of the embryo/fetus. Report 128. NCRP, Bethesda. http://www.ncrppublications.org/Reports/128
31. Berkovski V (1999) Radioiodine biokinetics in the mother and fetus. Part 1. Pregnant woman. In: Radiation and Thyroid Cancer. World Scientific Publishing. Publication No. EUR 18552 EN of the European Commission, pp 319–325
32. Berkovski V (1999b) Radioiodine biokinetics in the mother and fetus. Part 2. Fetus. In: Radiation and Thyroid Cancer. World Scientific Publishing. Publication No. EUR 18552 EN of the European Commission, pp 327–332
33. Anspaugh LR, Napier BA (2009) Minor parameters need for individual dose calculations, PNNL-18917. Pacific Northwest National Laboratory, Richland, WA

34. Berg GEB, Nyström EH, Jacobsson L, Lindberg S, Lindstedt RG, Mattsson S, Niklasson CA, Norén AH, Westphal OGA (1998) Radioiodine treatment of hyperthyroidism in a pregnant woman. J Nucl Med 39(2):357–361
35. ICRP (1992) Radiological protection in biomedical research ICRP Publication 62. Ann ICRP 22(3)
36. Ahlgren L, Ivarsson S, Johansson L, Mattsson S, Nosslin B (1985) Excretion of radionuclides in human breast milk after the administration of radiopharmaceuticals. J Nucl Med 26:1085–1090
37. Mountford PJ, Coakle AJ (1989) A review of the secretion of radioactivity in human breast milk: data, quantitative analysis and recommendations. Nucl Med Commun 10(1):15–27
38. Stabin MG, Breitz HB (2000) Breast milk excretion of radiopharmaceuticals: mechanisms, findings, and radiation dosimetry. J Nucl Med 41:863–873
39. Safety guide for radiation protection in nuclear medicine. Australian Radiation Protection and Nuclear Safety Agency, Radiation Protection Series Publication No. 14.2, 2008. ISBN 978-0-9805638-0-1, ISSN 1445-9760
40. Treves ST (ed) (2007) Pediatric nuclear medicine/PET, 3rd edn. Heidelberg, Springer. ISBN 038732321X
41. Gray L, Torreggiani W, O'Reilly G (2008) Paediatric diagnostic reference levels in nuclear medicine imaging in Ireland. Brit J Radiol 81:918–919
42. Ted Treves S, Davis RT, Fahey FH (2008) Administered radiopharmaceutical doses in children: a survey of 13 pediatric hospitals in North America. J Nucl Med 49:1024–1027
43. Lassmann M, Biassoni L, Monsieurs M, Franzius C, Jacobs F, Dosimetry EANM, Committees P (2007) The new EANM paediatric dosage card. Eur J Nucl Med Mol Imag 34(5):796–798
44. Lassmann M, Biassoni L, Monsieurs M, Franzius C (2008) EANM dosimetry and paediatrics committees. The New EANM paediatric dosage card: additional notes with respect to F-18. Eur J Nucl Med Mol Imag 35(9):1666–1668
45. Gelfand MJ, Prisi MT, Ted Treves S (2011) Pediatric radiopharmaceutical administered doses: 2010 North America consensus guidelines. J Nucl Med 52(2):318–322

Part IV
Quality Assurance and Quality Control

Chapter 6
Quality Control of Gamma Cameras, SPECT/CT and PET/CT Units

Jenny Oddstig, David Minarik, and Mikael Gunnarsson

6.1 Quality Control of Nuclear Medicine Imaging Equipment

6.1.1 Background

The image of the activity distribution reproduced by the gamma camera is influenced by the properties of the camera. These properties are affected by a series of parameters of which the user ought to have control over and for some of them also influence on. The objective at every examination is to produce an image of the activity distribution in an optimal way and with full control of the gamma camera properties. Regular quality control of the gamma camera is of decisive importance for the reliability of the result. Non-optimized conditions can even give rise to an incorrect diagnosis.

Example 1: The scintigraphy showed a local uptake in the skeleton. The image taken after decontamination of the collimator showed a normal scintigraphy without any locally increased uptake in the skeleton. The first registered images are a result of a contamination on the collimator which should have been discovered in a quality control of the camera. The right image is registered after decontamination of the collimator (Fig. 6.1) [1].

Example 2: A broken amplifier to a photomultiplier tube caused a 6-cm artifact in a pulmonary perfusion scan. The artifact was interpreted as a suspected pulmonary embolism. The physician decided to do a ventilation scan of the patient, and an

J. Oddstig (✉)
Radiation Physics, Skåne University Hospital Lund, Getingevägen 4, 221 85 Lund, Sweden
e-mail: jenny.oddstig@skane.se

D. Minarik • M. Gunnarsson
Radiation Physics, Skåne University Hospital Malmö, Malmö, Sweden

Medical Radiation Physics, Department of Clinical Sciences Malmö, Lund University, Malmö, Sweden

S. Mattsson and C. Hoeschen (eds.), *Radiation Protection in Nuclear Medicine*,
DOI 10.1007/978-3-642-31167-3_6, © Springer-Verlag Berlin Heidelberg 2013

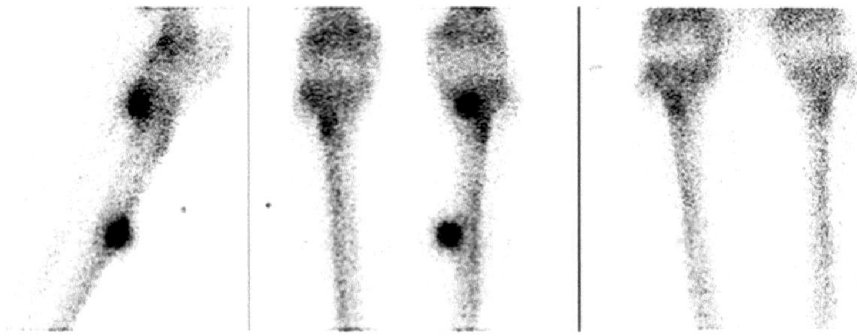

Fig. 6.1 Local uptake in a bone scintigram caused by contamination on the collimator; the right image is taken after decontamination

unnecessary activity administration of 100-MBq ^{99m}Tc-MAA was given. When the same artifact could be seen in another patient the next day, the staff realized that there was something wrong with the images. After a quality control of the gamma camera, the defect could be confirmed, and the pulmonary embolism ruled out after a new perfusion scan [2].

Regular quality controls are important to avoid this kind of mistakes. The gamma camera can, if there are deficiencies in its function, give false-positive as well as false-negative examination results. A quality control (QC) program is aimed to maintain a high performance of the equipment regardless of its age. The quality control starts with an initial *acceptance test*, which is based on a QC program to assess whether the equipment fulfills the specifications or not at installation. Routine QC measurements should then be made with regular intervals and also after major changes of components, updates by the manufacturer, or repairs. As it is essential to maintain long-term overall stability of performance, these measurements must be carefully specified, performed, recorded, and evaluated.

QC programs ought to be designed to follow international guidelines such as those from IAEA, EANM, AAPM, CEC (RP91, and soon RP106), IEC, and NEMA. These recommendations must though be considered in the light of any national guidelines and legislation, which must be followed. Also, the recommendations from the manufacturer must be taken into consideration. To follow the national legislation together with the recommendation from the manufacturers must be the minimum QC program for a nuclear medicine department.

A modern nuclear medicine department uses a number of different equipment for measurement of radiation, e.g., activity meters, thyroid uptake probes, gamma counting systems, radiation monitors, gamma cameras, positron-emission tomography (PET), CT scanners, etc. All clinically used equipment should be objects for quality control to ensure the correctness of the measured values. This chapter covers gamma cameras and SPECT/CT and PET/CT units.

6.1.2 Parameters Influencing the Image Quality

In nuclear medicine, there are no jointly agreed criteria for image quality like in X-ray imaging. Instead, it is a matter of the observer's subjective judgment of the image. The base for image quality is the possibility to detect differences in the uptake of radiopharmaceuticals in a lesion and in the surrounding tissue. An image of high quality is one that can reproduce this contrast in a way to secure a correct diagnosis.

However, there are several factors that will influence the image quality. Some of them are impossible to influence like the size of patient and organ localization. A large patient will increase the influence of scattered photons. Organs of interest will be overlapped by other tissues, increasing the background registration. Organ and patient movement will also decrease the image quality.

6.1.2.1 Uniformity

A homogenous irradiation of the detector should result in a uniform image. Good uniformity is necessary to get an image that accurately reflects the distribution of the radiopharmaceutical in the patient. One reason for nonuniformity can be a bad photomultiplier (PM) tube. The PM-tube gain can change with many environmental factors, such as room temperature, patient load, short-term or long-term radiation exposure, and time. An unbalanced PM tube degrades the image resolution and quality in both positron-emission tomography (PET) cameras and in gamma cameras. Other reasons for nonuniformity can be deficiencies in the detector or collimator, contamination, or incorrect settings of the energy window.

6.1.2.2 Energy Window

It is of importance to correctly center the pulse height window around the full-energy peak to get a high image quality. An incorrectly chosen energy window (against low energies) will detect a higher proportion of scattered photons and a smaller part of the photo peak, as is the case in Fig. 6.2. The more narrow energy window that can be used, the less Compton-scattered photons will be detected resulting in less noise in the image which increases both the resolution and the contrast. But a more narrow energy window gives also a lower count density.

6.1.2.3 Spatial Resolution

The spatial resolution is the ability to accurately resolve spatially separated radioactive sources from each other. There are several factors affecting the spatial resolution such as the collimator, distance between detector and radiation sources, scattering, attenuation, matrix size, and image processing.

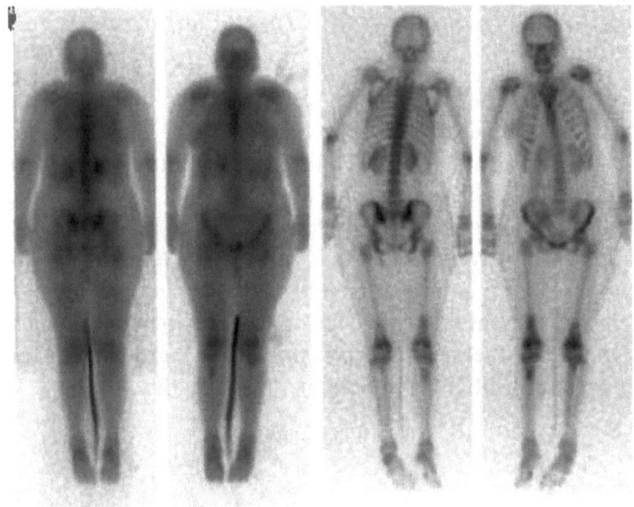

Fig. 6.2 Gamma camera images of a bone scintigraphy with ^{99m}Tc-EHDP measured in a ^{57}Co window (the energy window is positioned below the photo peak from ^{99m}Tc) to the left and in a ^{99m}Tc window (the energy window is symmetric located around the photo peak) to the right

If the user is aware of some of these factors, the image quality can easily be increased. By keeping the distance between the collimator and the patient as short as possible and selecting an optimal examination time, the resolution will be increased. An increase of matrix size, i.e., smaller pixels, will give a higher resolution but also a decrease in count density.

The choice of collimator also influences the spatial resolution. The longer and smaller the collimation holes are, the better will the resolution be, but on the expense of the sensitivity that will decrease with longer and smaller collimator holes. Also, the thickness of the walls between holes, the septum, affects the spatial resolution; penetration through the walls causes noise, but an unnecessary thick wall will decrease the resolution. The thickness of the septa should be suited for the photon energy used.

A fraction of the photons emitted in the lesion or organ of interest will be attenuated in the patient and never reach the detector. To get a correct image of the activity distribution, it is necessary to correct for this attenuation. The correction could be done with a mathematical model (often Chang model) or, as is more and more frequently done, with separate transmission measurement. The transmission measurement can be done with photons from a radioactive source (e.g., ^{153}Gd) or from an X-ray tube. The use of CT for attenuation correction has increased rapidly the last years. For more details, the reader is referred to Chap. 5 in Vol. 1 "Radiation Physics for Nuclear Medicine" in this series.

In modern gamma cameras, the system spatial resolution (FWHM) including both the detector and the collimator is around 10 mm. Corresponding value for a PET camera is 5 mm and for a CT 0.2 mm.

6.1.2.4 Contrast

Contrast is defined as the difference between object signal and the signal from the surrounding tissue. To achieve a high contrast it is important to have a suitable radiopharmaceutical, since the contrast depends on its biodistribution. With a high uptake in the tissue surrounding the lesion, it will be difficult to accomplish any high contrast. The contrast is also influenced by the parameters used in the image processing such as filter, image reconstruction algorithms, color scale, and threshold level.

6.1.2.5 Spatial Linearity

Failure in the spatial linearity correction will cause spatial distortion of the image with respect to the object. A poor linearity also gives reason to a poor uniformity of the image. The reasons for nonlinearity are the same as for nonuniformity.

6.1.2.6 Count Density in the Image

The fluctuations of count density in the flood phantom image are produced not only by the variation of the camera performance but also by statistical variations due to the stochastic nature of the radioactive decay. This statistical noise is reduced as the image count density is increased, with the result that non-stochastic variations in camera response are more accurately defined. Too few counts in the image may result in misinterpretation. An increase of the activity will increase the count density but also the radiation dose to the patient. Instead, a longer acquisition time could be used with the risk of artifacts due to patient and organ motions. Smaller matrix size (larger pixels) will also increase the count density but is on the expense of the spatial resolution that will be decreased with a decreased matrix size.

6.1.2.7 Image Processing

The processing of the image has also an influence on the image quality. The choice of filters, reconstruction algorithm, correction for scatter, and attenuation must be optimized for each type of investigation and camera—all to require a high image quality. Also, color scale and threshold level must be optimized to display the image in the best way.

6.1.3 Guidelines

6.1.3.1 The National Electrical Manufacturers Association

The National Electrical Manufacturers Association (NEMA) is a trade association for the electrical manufacturing industry in USA. The purpose of its publication is to provide a uniform criterion for the measurement and reporting of performance parameters by which a manufacturer may specify their device and, when doing so, refer to the NEMA standard. The standard does not establish any minimum performance level. By following the same standard as the manufacturer, it is possible for the purchaser to control that the equipment fulfills the specifications after installation [3–6].

Specific measurement equipment is required in order to accomplish the standard of the protocol. A majority of the measurements in the NEMA standard are done "intrinsically" (without collimator) and do not correspond to the clinical situation. Measurements complementing the NEMA protocol, corresponding more to the clinical situation, during the acceptance test are therefore recommended.

6.1.3.2 International Electrotechnical Commission

The International Electrotechnical Commission (IEC) is a nonprofit, nongovernmental standards organization that prepares and publishes international standards for all electrical, electronic, and related technologies. Today, the IEC is the world's leading international organization in its field, and its standards are adopted as national standards by the members [7–9].

6.1.3.3 International Atomic Energy Agency

The International Atomic Energy Agency (IAEA) has in 2009 published extensive proposals for quality assurance for both SPECT and PET cameras. The publications are free of charge and can be found at http://www.iaea.org. The extensive guidelines contain purpose of tests, material, procedures, data analysis, observations, interpretation of results, limits of acceptability, and conclusions for each part of the system needed to be tested. Test schedules are also listed with a proposal which tests to be included in the acceptance and reference test, respectively, and with which frequency the tests should be implemented in the clinical routine. To follow the IAEA guideline does not require any special phantoms or equipment and can be performed under simple conditions. The IAEA guideline includes tests under more clinical conditions than the NEMA and IEC tests and refers to the NEMA and IEC for control that the equipment fulfills the specifications after installation [10, 11].

IAEA has also published the "IAEA Quality Control Atlas for Scintillation Camera Systems," containing images of how different defects can influence the scintillation image and comments on how the effects could be corrected [12].

6.1.3.4 European Association of Nuclear Medicine

The European Association of Nuclear Medicine (EANM) is the organization of nuclear medicine professionals in Europe and represents also the field towards the European institutions. The EANM guideline consists of two parts: one for acceptance testing and one for routine quality controls [13, 14]. In both protocols, the purpose for the different tests is specified, but no proposal on how to carry out the test is given. Neither are any limits of acceptability specified. There is a frequency scheme of tests to do daily, monthly, and yearly. Like IAEA, EANM refers to the NEMA and IEC for the control that the equipment fulfills the specifications after an installation.

6.1.3.5 CEC: Radiation Protection 91 (RP91), Will Soon Be Replaced by RP106

The RP91 is a publication from the European Commission in the field of radiation protection with criteria for acceptability of radiological (including radiotherapy) and nuclear medicine installations. The purpose of the document is to specify a minimum standard of performance and find a minimum level when remedial actions need to be initiated. Values are specified for uniformity, sensitivity, geometry, linearity, etc., and when the measured value exceeds the specified value, this should be taken as an indication of remedial actions [15]. This publication will soon be replaced by an updated and more extensive new publication (RP106).

6.1.4 Types of Tests

6.1.4.1 Acceptance and Reference Tests

All nuclear medicine instruments must undergo acceptance tests after installation and before it is put into clinical use and before the final payment for the equipment. The acceptance tests are preformed to verify that the instrument performs according to the specifications, to test the clinical performance, and to create baseline data for following routine tests. Each instrument is supplied with a set of basic specifications produced by the manufacturer according to standard test procedures following a standard protocol such as the NEMA or IEC performance standards [3, 6–9]. By following the standard procedure in these protocols and with eventual support from the manufacture with special phantoms and software where necessary, the specifications can be verified. It is important to be aware of which standard protocol the manufacture used when measuring the basic specifications. Other guidelines like those from EANM and IAEA stress the importance to follow the NEMA and IEC to be able to compare the performance standard of the instrument with the specifications, but they also point out the lack of acceptance test of the clinical circumstance in the NEMA and IEC guidelines and recommends therefore further tests as a complement.

Acceptance testing is of concern for the maintenance staff, the manufacturer's agent, and the users of the instrument, and all should be involved. Acceptance testing is extremely important, as it can affect the whole life performance of the system. There is often a pressure to put new systems into clinical use directly after the company installation has ended, but it is important to reserve time to perform the acceptance test and investigate various upcoming issues. Adequate time at this stage may save many days of dealing with problems in the future.

6.1.4.2 Routine Tests

When the instrument has been accepted for clinical use, its performance needs to be tested routinely. A fundamental principle in the quality control of nuclear medicine instruments is that the quality control should be undertaken as an integral part of the routine work of the nuclear medicine department. The QC program should be designed to be sensitive to changes in the performance of the equipment but not as complicated as the acceptance test to perform. The routine test needs reference values to compare with, carried out at the time of the acceptance testing. All test results must be recorded and monitored for variations, and appropriate actions taken when deviations are observed.

The test frequencies for the equipment should be followed and adjusted accordingly to the observations of stability of the equipment and environmental stability, e.g., power support and room temperature. New equipment needs shorter time period between the test intervals, and the test frequency can be reduced only if the testing gives evidence that the system is stable. For older systems where the testing indicates a lack of stability, the testing frequency should be increased. Guidelines from different organizations like IAEA, EANM, and CEC (RP91, RP106) have different opinions about the frequency with which the routine test should be carried out and what they should contain. The recommendations from the manufacturers contain different test protocols and frequencies; different models from the same manufacturer have also different protocols. The recommendations for basic routine QC consist of test schedule with different frequencies for different tests, in most cases, divided in daily, weekly/monthly, and 6-monthly/yearly tests.

6.2 Quality Control Program for a Gamma Camera

A comprehensive agreement can be seen when studying the different guidelines. They include more or less the same tests for the routine control, and they agree for which tests to perform daily, but there is some discrepancy in the frequency schedule for weekly, monthly, and half-yearly tests. The guidelines from IAEA, EANM, etc., include a more extensive acceptance program than the NEMA and IEC publications. The NEMA and IEC did never intend to cover the clinical situation; they were set up to give reference values for specifications, and the other guidelines refer to NEMA and IEC for the control of specifications after installation.

A summary of the recommendations in the different guidelines and recommended test frequencies can be found below.

6.2.1 *Daily*

The daily routine at a nuclear department should include a control of the gamma camera before the first patient measurement starts. A physical inspection of the gamma camera is a good start to check collimator and detector for any damage together with a collimator touch pad and gantry emergency stop test. If any collimator damage is detected or suspected, an extrinsic high-count uniformity test should be performed. Also, a verification of the background count rate should be performed to ensure that there is no contamination.

6.2.1.1 Energy Window

A correct setting of the energy window is central for the image quality (see Fig. 6.2). Therefore, the energy window settings should be checked every day. Place a ^{99m}Tc source under the collimator. The source can be a syringe, a vial, or if necessary the patient. Analyze the energy spectrum. Accept the proposed energy window if the settings are correct or set a new proper window. Repeat the test for each radionuclide clinically used during the day. All photo peaks must be properly centered.

6.2.1.2 Flood-Field Uniformity

The flood-field uniformity can be tested intrinsically (without collimator) or extrinsically (with collimator). For the daily test of the uniformity, there is a need to sample four million counts or more to avoid influences form the counting statistics on the system uniformity. Position a point source of ^{99m}Tc (or ^{57}Co) at a large distance (5$\times$ the useful field of view) from the detector or more easily place a flood source on the collimator and sample. Analyze the image for inhomogeneity by either visual inspection or digital analysis (following the manufacturer's recommendations). If the intrinsic alternative is selected, each collimator must be checked periodically by an extrinsic uniformity test (preferably with a high-count acquisition). If the uniformity does not fullfill the recommendation, it is necessary to do a tuning of the PM tubes and sample a new flood-field uniformity matrix.

The planar uniformity of a scintillation camera should be better than 4% if the camera is used for SPECT.

6.2.2 *Weekly or Monthly*

6.2.2.1 Flood-Field Uniformity Intrinsic or Extrinsic

Quantitative analysis of the inhomogeneities in the image with a high number of counts to monitor trends in uniformity is recommended to be performed weekly or with a stable system monthly. The performance is the same as for the daily test but with a higher number of counts, preferably more than ten million counts. The quantitative analysis should be done for both integral uniformity and differential uniformity for the central and useful field of view based on maximum and minimum number of counts in an area. If an intrinsic method is selected, a high-count extrinsic measurement is also required routinely (half-yearly/yearly) to have a good control of the collimator.

6.2.2.2 Center of Rotation Offset and Alignment

Gamma cameras used for SPECT have to be routinely controlled for the offset and alignment of its center of rotation (COR). A control that the mechanical and electronically COR are aligned within the acceptability in *X*- and *Y*-direction has to be performed weekly and for a more stable system monthly. The test should be made for all collimators and for each detector configuration used in SPECT studies. Five small ^{99m}Tc point sources are placed along the axis of rotation, about 2 and 10 cm of the axis of rotation, and within 2 and 10 cm of the center of the field of view, respectively. Perform a normal tomographic acquisition collecting about 10,000 counts at every angular position, and compare the point source position to the expected path by calculating the offset from the COR in the reconstructed images [10].

6.2.2.3 System Sensitivity

The sensitivity, recorded counts per second (cps) per MBq, can be calculated from the extrinsic flood-field uniformity control. IAEA and EANM recommend a control of the sensitivity both for the daily low-count uniformity check and weekly/monthly from the high-count uniformity control [10, 14]. The sensitivity depends on the collimator properties and is higher for low-energy collimators than for high-energy collimator and higher in a low-resolution collimator than in a high-resolution collimator. Changes in the sensitivity might indicate incorrect energy window or could result from impaired energy resolution or nonuniformity in flood-field response.

6.2.2.4 Spatial Resolution and Spatial Linearity

A visual inspection of the spatial resolution and spatial linearity can easily be done by positioning a four-quadrant bar pattern on a flood source of ^{99m}Tc or ^{57}Co and do a short acquisition. It is important to have a phantom that matches the camera resolution, the smallest distinguishable separation should be (FWHM/1.75) to be able to evaluate the camera performance. Alternatively, a line source could be measured and the FWHM calculated from a line profile drawn over the line source. The spatial resolution and spatial linearity can be tested intrinsically or extrinsically. IAEA recommends weekly tests of resolution and linearity, while EAMN proposes a test frequency of 6 months [10, 14].

6.2.3 *Annually*

6.2.3.1 Intrinsic Spatial Resolution and Spatial Linearity

Control of the intrinsic spatial resolution is only necessary to do if the weekly/ monthly spatial resolution is tested extrinsically. The intrinsic test of the spatial resolution is done in the same way as described above without collimators.

6.2.3.2 Intrinsic Uniformity for Other Radionuclides

The intrinsic flood-field uniformity for all other radionuclides used in for imaging in clinical routine than ^{99m}Tc should be tested routinely. The measurements are done the same way as earlier described.

6.2.3.3 Tomographic Uniformity

The tomographic uniformity can be controlled by measuring a homogenously filled uniformity phantom. The test should be performed after making sure that the planar uniformity is correct and calibration for uniformity correction has been done. Perform a tomographic acquisition with the parameters used clinically, and perform uniformity corrections, attenuation, and scatter correction as recommended by the manufactures in the reconstruction. Inspect and measure the uniformity at various transaxial positions.

6.2.3.4 Tomographic Resolution

The tomographic resolution is measured to ensure that the resolution is not degraded by the tomographic acquisition or the reconstruction. Small point sources are placed in the same way as for the COR measurement free in air, and an acquisition is made in the same way as for the COR. After the tomographic acquisition, a static acquisition at the home position is performed. Analyze the data by drawing a line profile over the point source in the static and reconstructed images to calculate the FWHM. Compare the FWHM for the reconstructed images horizontally and vertically and the static image [10]. The tomographic resolution with scatter should also be tested to get more clinical comparable data. Repeat the measurements above with the point sources placed in a phantom which can generate scatter [10].

6.2.3.5 Intrinsic Flood-Field Uniformity with Narrowed and Asymmetric Energy Window

Intrinsic measurement with narrowed and asymmetric energy window over the upper half of the photo peak and over the lower half of the photo peak, respectively, is sampled, and the images are visually compared. Areas of nonuniformities indicate poorly tuned PM tubes or lack of integrity in the optical coupling between the PM tubes and the light-guide assembly [10].

6.2.3.6 Pixel Size

The pixel size can be measured with two point sources placed at the collimator with known distance. By measuring the distance in the image and dividing it with the distance in reality, the pixel size will be known. The pixel size is important if quantitative measurements should be done.

6.2.3.7 Overall System Performance

The overall system performance is tested with a performance phantom (e.g., Jaszczak or Carlson phantom) containing hot and cold regions to verify that the system performs adequately in high-count studies. The phantom should contain about 400 MBq ^{99m}Tc. It is also desirable that the phantom has some possibility for estimation of the resolution, e.g., rods. Perform a tomographic acquisition of the phantom using clinically relevant parameters, and reconstruct the data with the clinically used protocol. Review the transaxial slice looking for artifacts. Monitor the uniformity in slices without hot and cold regions and the contrast resolution in slices with rods or cold spheres [10, 14].

6.2.4 Acceptance and Reference Test

There is a good agreement between different guidelines of which tests to include in the acceptance test of new equipment [6, 10, 13]. A summary of the acceptance test from the guidelines is listed below. Most of these tests are described in the section for daily, monthly, and yearly controls and can be performed the same way as proposed in these sections.

The reference test should consist of all tests the clinic has intention to include in the daily, monthly, and yearly control program to create reference values for these controls.

6.3 Quality Control of PET Cameras

Most manufacturers of PET and PET/CT equipment recommend procedures for routine quality control, and each manufacturer defines procedures that generally are specific for their own products. It is recommended that users follow these recommendations, but in the absence of such recommendations, a minimum QC protocol that all owners of PET or PET/CT equipment should carry out is described below. Compared with stand-alone PET scanners, PET/CT systems require monitoring of additional parameters related to the conjunctional performance of the CT scanner with the PET scanner, i.e., the co-registration of CT and PET data and the accuracy of CT-based attenuation correction. The attenuation correction accuracy depends on the general performance of the CT, and a description of this is found in the CT quality assurance section. A summary of the minimum standard for PET QC from IAEA guidelines [11] can be found below.

6.3.1 Daily

The aim of the daily quality control is to assess the stability of the detector system and to discover eventual changes in performance before it starts to impact the image quality. As for a gamma camera, it is good to start with a visual inspection of the gantry and patient handling system to ensure that mechanical defects do not compromise the safety of the patient and staff. It is recommended that these daily QC procedures are carried out by the technologist normally operating the system.

A test of the detector modules is recommended to be performed daily. Depending on whether the PET system runs in a 2D or 3D acquisition mode, different sources could be used, such as ^{22}Na point or rod sources. The manufacturers of PET systems today supply recommendations of what type of source to be used and how the daily QC acquisition should be performed. The acquired sinograms should be visually or automatically inspected for the presence of pronounced diagonal streak artifacts. Most manufacturers have automated daily QC tools for analyzing and reporting the QC procedure. It is important to store the calculated primary parameters in order to detect any trends. The suggested tolerances of these parameters are usually supplied by the manufacturer. If a parameter falls outside the allowed tolerance or if artifacts in the sinograms or reconstructed images are present, normalization or calibration of the detector system should be considered.

IAEA also suggests that a coincidence timing resolution test is performed daily if the scanner can run in a time of flight mode. The test is accomplished by measuring a point source of ^{22}Na or other radionuclide as recommended by the manufacturer, within some scattering media. The timing resolution can then be estimated using a procedure supplied by the manufacturer. The timing resolution should be stable, and if the tolerance is exceeded, the test should be repeated. A recalibration of the system by appropriate service personnel should then be considered if the timing resolution still is outside of the tolerance level.

6.3.2 Monthly or Quarterly

IAEA recommends that a normalization measurement is performed monthly. The sensitivity of a particular line of response (LOR) is affected both by the geometry of the camera and the LOR position. Also, the block detectors themselves vary in efficiency, the PMT gains are not all exactly the same and may vary with time, and the scintillation crystals are not all identical. The normalization correction compensates for these variations in efficiency in each LOR. All manufacturers have a standard procedure for acquiring normalization data. Depending on the manufacturer and the acquisition mode, the normalization data can be acquired by using different sources and phantoms, such as a rotating ^{68}Ge line source, a uniform cylindrical ^{68}Ge phantom centered in the FOV of the scanner or a rotating ^{137}Cs point source. A visual inspection of the acquired normalization sinograms should be made, and if no problems are observed, the new normalization data should be stored as is established by the manufacturer. Recalibration of the system should be considered if problems with the normalization data exist, and if the problem persists, the manufacturer should be notified and maintenance scheduled.

IAEA also recommends that a radioactivity concentration verification/calibration is performed monthly. The aim of the verification test is to verify that the efficiency calibration data used to correct the acquired data is adequate. The calibration data are used to calculate activity concentration and standardized uptake values (SUV). Inaccurate calibration data will compromise accurate image-based activity quantitation. It is also recommended that new calibration data is collected at least quarterly. All manufacturers have established protocols for the calibration measurement. Generally, a cylindrical fillable phantom with known dimensions could be used, filled preferably with ^{18}F with known activity concentration. The verification is performed by measuring this phantom, and if the weight of the phantom and activity concentration is correctly typed into the measurement protocol, the mean SUV in a ROI in the reconstructed image should then be 1. Regarding the acquisition of calibration data, the instructions from the manufacturer should be followed.

It is recommended that a test of the uniformity is performed at least quarterly. The uniformity of a reconstructed image is a measure of the system response to a homogeneous distribution in both the transverse and axial FOV. If there is no homogeneity test protocol supplied by the manufacturer, it is recommended that the NEMA NU2 1994 [20] protocol is used.

Compared with stand-alone PET scanners, PET/CT systems require calibration and verification of the co-registration of PET and CT data. It is recommended that an offset calibration is performed quarterly or whenever the PET and CT gantries are separated for service. The purpose of the offset calibration is to ascertain that acquired PET and CT images co-register correctly. Incorrect offset values will result in misregistration errors in fused images and attenuation artifacts due to the use of misregistered attenuation correction maps. All manufacturers should have standardized protocols for offset calibration. Verification can be performed with any point or line source that are visible in both systems, e.g., a point source of ^{18}F mixed with a small amount of CT contrast agent. A visual inspection is generally enough to detect any offset between the PET and CT data.

6.3.3 Annually

It should be considered to monitor the consistency of the image quality parameters such as spatial resolution and contrast once every year. This could be performed using rod sources and nonuniform phantoms with hot and cold sphere. Details of these measurements are described in NEMA NU 2007 [3].

6.4 Quality Control of the CT Scanner

The aim of performing quality control of the CT part of a SPECT/CT and a PET/CT unit is to secure that the diagnostic outcome is correct. As mentioned earlier in this chapter, it is important to follow the recommendations of each manufacturer for the QA procedures. Each manufacturer delivers QA phantoms with the equipment, and most of the recommended procedures can be performed using these phantoms. An advantage of using these phantoms is that the specifications of the modality often are validated by these phantoms and it is easy to perform constancy control measurements based on the values from the manufacturer.

The recommendations from the organizations differ somewhat regarding how often the procedure should be performed; a summarization of guidelines for CT equipment can be found below.

6.4.1 *Daily*

The aim of the daily quality control is to assess the stability of the detector system and to discover eventual changes in performance before it starts to impact the image quality. It is recommended that these daily QC procedures are carried out by the technologist normally operating the system.

The daily QC routine recommended by most manufacturer and organizations includes a control of the CT before the first patient measurement starts. Starting the CT system always includes a tube warm-up and a calibration in air.

6.4.1.1 Tube Warm-Up

During the warm-up procedure, the tube heating in the X-ray tube is carefully increased in order to prevent damage to the tube and eliminate the potential for an arc to occur. The procedure includes a series of exposures made at incrementing tube voltage.

6.4.1.2 Calibration in Air

The procedure includes a series of exposures at varying techniques in order to normalize the detector response using air as the attenuating media. These scans essentially adjust the detector gains to achieve a uniform response.

Further procedures recommended by AAPM [16] that can be performed on a daily basis are *CT number accuracy of water, image noise, image uniformity,* and *artifacts.* These are the most critical tests since they are sensitive to a wide range of CT scanner problems. An advantage of these tests is that the measurements can be performed by using a simple water phantom, and the data for the tests are all achieved from the same phantom scan. The analysis is performed by visually inspection, looking for presence of artifacts caused by nonuniformity in detector response or putting a ROI in the image and analyzing the distribution of Hounsfield units (HU) in the homogeneous material. IAEA [17] also recommends that *CT laser alignment, tabletop alignment, positional accuracy, and scout scan accuracy* are tested on a daily basis if the modality is used for radiation treatment purposes.

6.4.2 *Monthly/Semiannually*

More complete versions of the tests described above should be performed on a monthly/semiannual basis. It is recommended that these QC procedures are carried out by the responsible medical physicist. Further procedures that should be tested are mentioned below.

6.4.2.1 CT Laser Alignment, Tabletop Alignment

The aim of these tests is to ensure that the gantry lasers and room alignment lasers (if the modality is used for radiation treatment purposes) are properly aligned with the CT gantry and the position of the table. The tabletop alignment test is to ensure that the movement of the table according to the position indicators is accurate and reproducible. The test is performed by putting 70 and 140 kg weights on the tabletop and moving the table vertically. The tolerance criterion given by IAEA [17] is for both procedures ±1 mm.

IAEA [17] recommendations are that these QA procedures are performed at least monthly and whenever the alignment lasers are serviced. If, however, images are to be used for radiation therapy treatment planning, these tests should be carried out daily or at least on those days prior to using the system for treatment planning purposes.

6.4.2.2 Uniformity and Noise

It is recommended that a test of the uniformity and noise level is performed at least monthly. Most manufacturers have service protocols including uniformity/noise tests that are performed on the QA phantom that is delivered with the CT. The uniformity (the difference between the average CT number at the center and that at the periphery) should be within ±10 HU (CT numbers) according to IAEA [17]. The corresponding value of the image noise (the standard deviation divided by the average of the CT numbers of water in the center of the phantom) should be $<\pm 10\%$.

6.4.2.3 Computed Tomography Number and Electron Density Accuracy

The aims of these tests are to ensure that the CT numbers for different materials are within the appropriate limits. A phantom containing a variety of materials with a wide range of CT numbers is scanned and compared with standard values from the manufacturer of the phantom. The materials required consist of a phantom filled with water and with regions of different densities, typically including polyethylene, PMMA, nylon, Teflon, and air. The tolerance criteria given by IAEA [17] is ±5 HU for CT numbers specified by manufacturer of the phantom.

IAEA [17] and AAPM [16] recommend that the test is performed monthly to semiannually and for radiation therapy applications daily or before patient scans. IAEA [17] also stresses that the procedures must also be carried out whenever an X-ray tube is replaced, on system calibration, generator maintenance, software changes or upgrades, or on any other invasive service that may affect the CT number accuracy, uniformity, noise, or image.

6.4.2.4 High-Contrast Modulation

The aim of this test is to ensure that images with good modulation of high-contrast objects, i.e., small details, will be imaged with good fidelity. Some CT scanners are capable of calculating resolution limits, MTFs, or point spread functions automatically by scanning a phantom containing a resolution phantom or thin wire. There is also a commercial software available that do these calculations automatically after analyzing the DICOM information from the CT scan. The tolerance criterion given by IAEA [17] is ±15%.

6.4.3 Annually

In addition to the QC tests described above, there are additional QC tests that should be carried out on an annual basis by the responsible medical physicist.

6.4.3.1 Radiation Dose

In order to ensure the radiation dose from the CT, the computed tomography dose index (CTDI) is measured. CTDI represents the average absorbed dose along the z-axis from a series of contiguous irradiations. The most commonly used index is CTDI_{100}; this measurement requires the use of specified CTDI phantoms composed of PMMA with diameters of 16 and 32 cm. The dose is integrated over a length of 100 mm by using a pencil ionization chamber with an active length of 100 mm:

$$\mathrm{CTDI}_{100} = \frac{1}{nT}\int_{-50\ \mathrm{mm}}^{50\ \mathrm{mm}} D(z)\mathrm{d}z, \tag{6.1}$$

where n is the number of slices per rotation, T is the nominal slice thickness, and $D(z)$ is the dose absorption distribution along the z-axis, i.e., the dose profile. To account for the spatial variation of the dose in the scan plane (x, y), a weighted dose index (CTDI_w) was introduced [18]:

$$\mathrm{CTDI}_w = \frac{1}{3}\mathrm{CTDI}_{100(\mathrm{central})} + \frac{2}{3}\mathrm{CTDI}_{100(\mathrm{peripheral})}. \tag{6.2}$$

To take axial scan spacing into account, the CTDI by volume ($\mathrm{CTDI}_{\mathrm{vol}}$) was introduced: [19]:

$$\mathrm{CTDI}_{\mathrm{vol}} = \frac{\mathrm{CTDI}_w}{\mathrm{pitch}}, \tag{6.3}$$

where pitch is defined as the ratio between the table transportation per rotation and the total nominal slice width. $CTDI_{vol}$ is expressed in mGy and is displayed on the CT consoles. The $CTDI_{vol}$ is a measure of the radiation output of a CT scanner and represents an estimation of the average radiation dose within the irradiated volume of an object of similar attenuation to the CTDI phantom.

The recommendations from IAEA [17] are that CTDI measurements should be performed annually or after major services, e.g., replacement of X-ray tube. The tolerance criteria given by IAEA [17] are ±20% of manufacturer's specifications.

6.4.3.2 Half-Value Layer

The aim of this test is to ensure that the CT equipment meets the specifications from the manufacturer regarding regulations of filtration of the CT beam.

6.4.3.3 *Z*-Axis Characteristics

The aim of this test is to ensure the *z*-axis characteristics. The slice sensitivity profile (SSP) is usually measured by using the phantom delivered with the CT by the manufacturer. The SSP section of the phantom contains of a bead phantom, and after subtracting the background in the CT image, one can calculate the FWHM. The recommendations from AAPM [16] are that the shape of the SSP obtained with a bead phantom and reconstructed at one-tenth of the collimation should be inspected to ensure that it is a smooth curve and that the FWHM meets the manufacturer's specification.

References

1. Carlsson S, Svensson S-E (2007) Nuklearmedicin. Swedish Society for Nuclear Medicine, Sweden, p 1–276. http://www.sfnm.se/Nuklearmedicin_SC_SES.pdf (in Swedish)
2. Swedish Radiation Safety Authority (2009) http://www.stralsakerhetsmyndigheten.se/Yrkesverksam/Vard/Aterkoppling-av-missoden/Nuclearmedicin/Tekniskt-fel-hos-gammakamera-ledde-till-felaktig-diagnos/
3. Association of Electrical Equipment and Medical Imaging Manufacturers (NEMA) (2007) Performance measurements of positron emission tomographs. NEMA NU2
4. Association of Electrical Equipment and Medical Imaging Manufacturers (NEMA) (2004) Performance measurements and quality control. Guidelines for non-imaging intraoperative gamma probes. NEMA NU3
5. Association of Electrical Equipment and Medical Imaging Manufacturers (NEMA) (2008) Performance measurement of small animal Positron Emission Tomographs. NEMA NU4
6. Association of Electrical Equipment and Medical Imaging Manufacturers (NEMA) (2001) Performance measurement of scintillation cameras. NEMA NU 1

7. International Electrotechnical Commission (2005) Medical electrical equipment—characteristics and test conditions of radionuclide imaging devices—anger type gamma cameras. IEC 60789—(Ort: Förlag) sidor
8. International Electrotechnical Commission (2005) Radionuclide imaging devices—characteristics and test conditions—Part 2: Single photon emission computed tomographs. IEC 61675-2
9. International Electrotechnical Commission (1998) IEC 61675-3 Radionuclide imaging devices—Characteristics and test conditions—Part 3: Gamma camera based whole body imaging systems
10. International Atomic Energy Agency (2009) Quality assurance for SPECT systems
11. International Atomic Energy Agency (2009) Quality assurance for PET and PET/CT systems. (Ort: Förlag) sidor
12. International Atomic Energy Agency (2003) Quality control atlas for scintillation camera system. (Ort: Förlag) sidor
13. Busemann Sokole E, Britten A, On behalf of the EANM Physics Committee (2010) Acceptance testing for nuclear medicine instrumentation. Eur J Nucl Med Mol Imag 37:672–681
14. Busemann Sokole E, Britten A, On Behalf of the EANM Physics Committee, With Contribution from the EANM Working Group on Nuclear Medicine Instrumentation Quality Control, Georgosopoulou ML, Tindale W, Klett R (2010) Routine quality control recommendations for nuclear medicine instrumentation. Eur J Nucl Med Mol Imag 37:662–671
15. European Commission (1997) Radiation protection 91. Criteria for acceptability of radiological (including radiotherapy) and nuclear medicine installations
16. Leitz W, Axelsson B, Szendrö G (1995) Computed tomography dose assessment—a practical approach. Radiat Prot Dosim 57:377–380
17. Association of Electrical Equipment and Medical Imaging Manufacturers (NEMA) (1994) Performance measurements of positron emission tomographs. NEMA NU2
18. American Association of Physicists in Medicine (2002) Quality control in diagnostic radiology. AAPM report no. 74. Medical Physics Publishing, Madison sidor
19. International Atomic Energy Agency (2009) Quality assurance for PET and PET/CT systems. IAEA human health series no. 1. IAEA, Vienna
20. International Electrotechnical Commission (IEC) (2009) Medical electrical equipment. Part 2–44: particular requirements for the basic safety and essential performance of X-ray equipment for computed tomography. IEC 60601-2-44, 3rd edn. IEC, Geneva

Part V
Occupational Exposure

Chapter 7
Occupational Exposure: With Special Reference to Skin Doses in Hands and Fingers

Adela Carnicer, Mercè Ginjaume, Marta Sans-Merce, Laurent Donadille, Ilona Barth, and Filip Vanhavere

7.1 Introduction

Nuclear medicine (NM) is the medical specialty that is associated with all uses of unsealed radioactive sources for diagnosis and treatment of disease. In diagnostics, technological advances have led to the fast spread of both the conventional and new imaging techniques such as single-photon emission computed tomography (SPECT) and positron-emission tomography (PET). As a consequence, radiopharmaceuticals are increasingly used, thus resulting in a rise of workload in radiopharmacy units and nuclear medicine departments. On the other hand, new therapy procedures with unsealed radionuclides are also gaining increasing importance. Pure beta-emitters or mixed beta-gamma radionuclides are particularly suitable for therapy applications with typically high activities required to fulfill the therapeutic effect.

Radiation protection of workers is an important issue in NM since, firstly, high radionuclide activities are needed, from few tens to several thousands of MBq; secondly, the procedures require the handling of radiopharmaceuticals at contact or very close to the extremities (hands, fingers); and, thirdly, often pure beta-emitters

A. Carnicer • M. Ginjaume (✉)
Universitat Politecnica de Catalunya (UPC), Diagonal 647, 08028 Barcelona, Spain
e-mail: merce.ginjaume@upc.edu

M. Sans-Merce
Institut de Radiophysique, University Hospital Center (CHUV), University of Lausanne, Rue du Grand-Pré 1, 1007 Lausanne, Switzerland

L. Donadille
Institut de Radioprotection et de Sûreté Nucléaire (IRSN), 31 Av. de la Division Leclerc, 92260 Fontenay-aux-Roses, France

I. Barth
Bundesamt für Strahlenschutz (BfS), Koepenicker Allee 102-130, 10318 Berlin, Germany

F. Vanhavere
Belgian Nuclear Research Centre (SCK-CEN), Boeretang 200, 02400 Mol, Belgium

S. Mattsson and C. Hoeschen (eds.), *Radiation Protection in Nuclear Medicine*,
DOI 10.1007/978-3-642-31167-3_7, © Springer-Verlag Berlin Heidelberg 2013

and mixed photon/beta-emitters are used. NM workers are thus potentially exposed to external radiation and to internal contamination in case of accidental intake. If adequate protocols are used, in general, contamination leads to negligible exposure to staff. External whole-body exposures for nuclear medicine staff are coming mostly from the patient contribution, in particular in PET procedures, but the annual effective dose is usually low (2–3 mSv for gamma procedures, around 6 mSv for PET). However, the exposure of the extremities during preparation and administration of radiopharmaceuticals can be high. The hands remain often unprotected, and thus, fingertips can receive high doses which are likely to exceed the dose limit for extremities whenever the level of radiation protection is insufficient or the workload is too high.

The works of Vanhavere et al. [1] and Donadille et al. [2] highlighted the fact that the radiation protection of workers in NM presented open issues that were not yet satisfactorily addressed. Thus, from January 2008 up to February 2011, the collaborative project, *Optimization of Radiation Protection of Medical Staff*, ORAMED, was set up and funded within the European Atomic Energy Community's Seventh Framework Programme (http://www.oramed-fp7.eu) with the aim of overcoming the problems previously identified. In particular, one of the working groups in ORAMED, WP4, aimed at the study of extremity dosimetry within NM. Three main objectives were proposed:

- To address the lack of knowledge on skin dose distribution and maximum skin dose to the hands
- To optimize routine monitoring of extremity dosimetry in order to assess skin dose as close as possible to maximum skin dose
- To set up the conditions and requirements necessary to ensure an acceptable level of radiation protection

This chapter first gives an overview of basic concepts, regulation, and problems associated with occupational monitoring in nuclear medicine. It then presents the methodology and main results of the ORAMED project in the field of extremity dosimetry of nuclear medicine staff. Finally, some recommendations to improve radiation protection in occupational exposure in nuclear medicine are proposed.

7.2 Occupational Monitoring: Basic Concepts, Regulation, and Practical Considerations

The main objectives of occupational monitoring are to provide a basis for estimation of the actual radiation exposure of workers and to demonstrate compliance with legal requirements. It is also useful to optimize operating procedures, to increase awareness of risk, and to motivate workers to reduce their own exposure. The limitation of dose for occupationally exposed workers to ionizing radiation is regulated by National and International Authorities. Regulations are based on the recommendations of the International Commission of Radiological Protection (ICRP) [3, 4] and the International Commission of Radiation Units and Measurements (ICRU) [5, 6]. In Europe, the Council Directive 96/29/Euratom [7] establishes the basic safety standards for the health protection of the general public and workers against the dangers of ionizing radiation. This directive is based on ICRP Publication 60 [3] and is now under revision [8]. The new version introduces recent scientific findings and recommendations, such as the 2007 recommendations of the ICRP [4].

Monitoring of internal exposure for nuclear medicine workers requires frequent measurements due to the short physical half-lives of most radionuclides used in this field. Baechler et al. [9] describe a protocol used in Switzerland to perform screening measurements of NM workers at the workplace to detect whether potential intake has occurred. The intakes from ingestion and inhalation are usually negligible, provided that adequate protection measures are applied. However, when volatile radionuclides such as iodine are used, it is recommended to monitor the workplace conditions, in particular to control contamination levels in the air.

The operational quantity for external individual monitoring (see Chap. 1) is the personal dose equivalent, $H_p(d)$, which is the dose equivalent in ICRU soft tissue [10] at an appropriate depth, d, below a specified point on the human body. For the assessment of effective dose due to external exposure, $H_p(10)$ at a depth $d = 10$ mm is chosen, and for the assessment of skin, hands, and feet equivalent dose, the personal dose equivalent, $H_p(0.07)$, at a depth $d = 0.07$ mm is used. A depth $d = 3$ mm has been proposed for the case of monitoring the dose to the lens of the eye.

Operational quantities are measurable, and instruments for radiation monitoring of external exposure are calibrated in terms of these quantities. In routine monitoring, the values of these operational quantities are usually a conservative estimate of the protection quantities. $H_p(d)$ is usually measured with a whole-body dosemeter, worn on the anterior part of the chest. The individual dose monitoring is performed with a passive dosemeter, mainly with thermoluminescent detectors but also with photographic films and photoluminescence and optically stimulated luminescent detectors. Likewise for internal exposure, the effective dose due to external radiation is usually low. According to the results from the ESOREX project [11], in the medical field, in Europe, for the year 2000, 93% of monitored workers received an annual effective dose below 2 mSv and 99% below 5 mSv. In the training material on radiation protection in PET/CT [12], the IAEA provides some typical annual doses in nuclear medicine: around 1 mSv for radiochemists, below 6 mSv for PET/CT, and below 0.1 mSv for the other staff. Indeed, as mentioned earlier, the largest contribution to whole-body exposures for nuclear medicine staff is mostly due to ^{18}F-injected patients. Radiation exposure from CT during PET–CT procedures, which imply higher patient doses, can be neglected for staff, because of the beam geometry and the fact that technologists are usually outside the irradiation room. This topic is discussed further in Chap. 11.

In cases in which exposure is not homogeneous or is localized on a different part of the body, the whole-body monitoring has to be completed with additional dosemeters worn on the exposed zone. This is the case of workers involved in the preparation, labeling, or injection of radiopharmaceuticals. The monitoring of

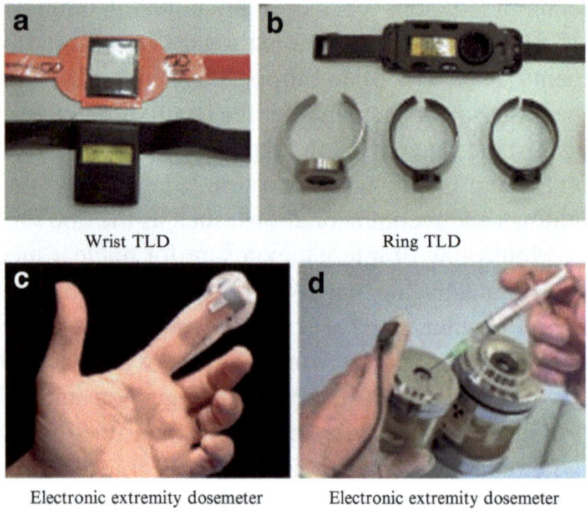

Fig. 7.1 Different types of extremity dosemeters

extremities and skin is recommended for workers that might receive an annual equivalent dose higher than 3/10th of the equivalent dose limits for hands and skin, namely, 150 mSv for hands or skin.

The most widely used dosemeters for the extremities are based on thermoluminescent detectors, placed in a holder that can be worn on the base of the finger or on the wrist. They are commonly known as *ring* and *wrist* dosemeters, respectively. Some dosemeters are specifically designed to be worn at the fingertips and also some electronic devices are available [13], but their use is much less frequent, mainly because they hinder the regular work. Figure 7.1 shows an example of the different types of extremity dosemeters.

7.3 Finger Doses for NM Workers

This paragraph provides information on finger doses for NM workers and guidance to monitor them, mainly based on the results of the ORAMED project.

7.3.1 Methodology

In order to determine the dose distribution across the hands and to supply information on reference dose levels for the most frequent NM procedures, an extensive measurement campaign was performed within the ORAMED project. It included 139 workers from 35 NM departments in seven European countries (Belgium, France, Germany, Italy, Slovakia, Spain, and Switzerland) representing the largest number of collected data on extremity dosimetry in NM up to now [14]. The experimental data were complemented with Monte Carlo (MC) simulations to better determine the main

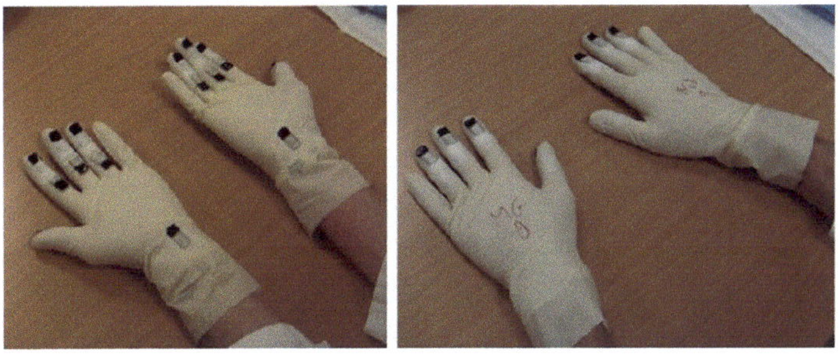

Fig. 7.2 Example of gloves used in the measurement campaign

parameters that influence extremity exposure, the effectiveness of different radiation protection measures and the degree of variability that could be "intrinsically related" to each monitored procedure. Details on the Monte Carlo protocol and results are described by Ferrari et al. [15].

For the measurement campaign, a common protocol was established to be able to compare and evaluate the data from the different hospitals. In particular, it was agreed to use thin detectors (effective thickness below 10 mg cm^{-2}) for positron and beta-emitters [16]. The operational personal dose equivalent $H_p(0.07)$ was measured at 11 positions on each hand (Fig. 7.2), considering both the usually highest exposed areas (fingertips and fingernails) and the most practical and frequently used positions for routine monitoring (wrist and bases of the fingers). The most frequently employed radionuclides were considered, i.e., ^{99m}Tc and ^{18}F for diagnostic applications and ^{90}Y for therapy. Measurements were performed separately for each radionuclide and independently for preparation and administration. For each worker, a set of 4–5 measurements were taken, except for therapy, where this was not always achievable.

For the analysis, the measured doses were normalized to the activity defined according to the following criteria:

- For preparation:
  - For ^{99m}Tc, the activity withdrawn from the elution vial to prepare the radiopharmaceutical (this is less than the total eluted activity)
  - For ^{18}F, the activity in the mono or multidose vial
  - For ^{90}Y, the activity used for the preparation of the radiopharmaceutical
- For administration:
  - The total activity in the injection syringe

Then, the mean normalized dose in each monitored position was calculated for each worker and for each procedure. From these data, the distribution of the maximum normalized dose in the monitored workers is obtained $\langle H_p(0.07)_{max}/A \rangle$.

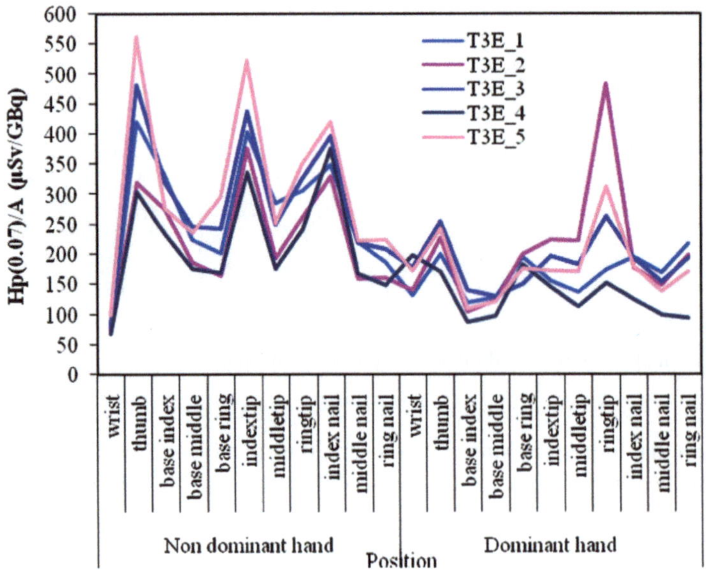

Fig. 7.3 Hand dose distribution for worker T3E, for preparation of ^{18}F. Each curve corresponds to individual sets of 20 TL readings

7.3.2 *Results and Discussion*

7.3.2.1 Hand Dose Distribution

As an example, Fig. 7.3 shows the normalized hand dose distribution during ^{18}F preparation. The worker is anonymously labeled T3E. The graph presents the set of five measurements of this worker; the uncertainty associated to each individual measurement ($k = 1$) is of the order of 15%. Although hand dose distribution varies between workers and techniques, general trends could be observed.

The tips of the fingers of both hands, especially the index and thumb, were identified to be the highest exposed positions. There is general agreement on this issue [17–21]. The least exposed positions were found to be the wrists, followed by the bases of the fingers. A clear trend was observed for the nondominant hand to be more exposed than the dominant hand, in particular for radionuclide preparation. However, this trend was strongly linked to individual working habits. In the literature, there is no consensus on which hand is the most exposed. The influence of individual working habits on the most exposed hand and position has also been pointed out in several works [17, 19, 20]. ICRP 106 [22], based on a thorough literature review, reports that the fingertips (especially index and thumb) of the dominant hand are the most exposed. For therapy, spatial dose inhomogeneity is usually much more pronounced, but generally also the same positions as for diagnostics were the most exposed. In most cases, the index tip of the nondominant hand is the most exposed specific position [23].

Table 7.1 Mean, median, maximum, and minimum values of $\langle H_p(0.07)_{max}/A\rangle$ of all monitored workers per procedure (*A* stands for administration and *P* for preparation) (adapted from Sans-Merce et al. [14])

	Maximum doses from all workers (mSv/GBq)			
	Mean	Median	Minimum	Maximum
P—^{99m}Tc	0.4	0.25	0.03	2.1
A—^{99m}Tc	0.2	0.12	0.01	0.9
P—^{18}F	1.2	0.83	0.1	4.4
A—^{18}F	0.9	0.64	0.1	4.1
P—^{90}Y Zevalin	11	9.5	1.2	44
A—^{90}Y Zevalin	5	2.9	1.0	12

7.3.2.2 Maximum Skin Dose to the Hands

Table 7.1 presents the range, median, and mean of $\langle H_p(0.07)_{max}/A\rangle$ overall monitored workers, classified per procedure. It is shown that preparation of radiopharmaceuticals involves higher finger doses per unit activity than administration because the procedures take longer time and there are more steps requiring manipulations of the vials and/or syringes with higher activities, some of them without a shield. Therapy procedures involve generally higher mean normalized skin dose to the hands than diagnostics. Within diagnostics, ^{18}F involves higher skin doses per unit activity than ^{99m}Tc because of the different dose rates at contact. Considering typical workloads, preparation of ^{18}F was found to be the most critical of the studied procedures, which is in agreement with other authors' findings [17, 18].

In Tables 7.2 and 7.3, ORAMED results are compared to earlier published data for diagnostics and therapy, respectively. For each referenced study, the tables show the number of monitored workers, the number of measurements per worker, and the values of $\langle H_p(0.07)_{max}/A\rangle$ (minimum, median, mean, and maximum). For diagnostics, the value (maximum or mean) and position [fingertips (tips) or base of fingers (ring)] of the reported doses are also provided. For therapy, all tabulated data correspond to maximum doses measured in the tip of the fingers. Rimpler et al. [24] data are given with and without outliers. Outliers correspond to cases in which radiation protection means were not standard, either because shielding was not used or because semiautomatic devices were used. Likewise, in Rimpler et al. [25], the authors reported some high doses which were considered outliers and were therefore not included in the calculation of the mean and median.

Unfortunately, not all available works could be included in the comparison because of major differences in the measurement methodologies (type of detectors, radionuclides, procedures, etc.) or in the expression of the results (e.g., doses not normalized to the manipulated activity) or because many details were omitted. Even after the selection of studies, comparison must be performed with care since, generally, some parameters differ to a certain extent from work to work. In spite of the large range of data, there is good agreement on the relative exposure for the considered procedures and on the position of the maximum exposure, the tips of the fingers.

Table 7.2 Comparison of values of hand skin dose in NM diagnostics in several published works (adapted from Carnicer et al. [23])

Procedure	References	N workers	Measurements/ worker	Value (position)	$\langle H_p(0.07)_{max}/A \rangle$ (μSv/GBq) Min	Median	Mean	Max
^{99m}Tc administration	Carnicer et al. [23]	32	4–5	Max (tip)	10	120	230	950
	Tandon et al. [37][a,c]	54	1–2	Mean (ring)	5	46	175	999
	Covens et al. [36][a,b]	5	n.s.	Max (tip)	40	49	50	60
^{99m}Tc preparation	Carnicer et al. [23]	36	4–5	Max (tip)	30	250	430	2,060
	Tandon et al. [37][a]	54	1–2	Mean (ring)	2	46	113	361
	Wrzesién et al. [35]	13	3–4[a]	Max (tip)	30	–	260	2,000
	Covens et al. [36][a,b]	2	n.s.	Max (tip)	20	65	65	110
	Leide-Svegborn [38][b]	3	3–7	Max (tip)	20	29	57	121
^{18}F administration	Carnicer et al. [23]	30	4–5	Max (tip)	140	640	930	4,110
	Tandon et al. [37][a]	3	1–2	Mean (ring)	155	218	232	324
	Covens et al. [36][a,b]	5	n.s.	Max (tip)	210	320	321	530
	Covens et al. [18][b]	8	5[a]	Max (tip)	200	280	350	750
	Covens et al. [18][b,d]	8	2–3[a]	Max (tip)	3.5	10	11	30
^{18}F preparation	Carnicer et al. [23]	30	4–5	Max (tip)	100	830	1,200	4,430
	Tandon et al. [37][a]	3	1–2	Mean (ring)	65	87	83	98
	Covens et al. [36][a,b]	2	n.s.	Max (tip)	290	570	570	850
	Covens et al. [18][b]	2	25[a]	Max (tip)	90	320	500	1,300
	Covens et al. [18][b,d]	2	10[a]	Max (tip)	4.5	9	10	18

n.s. Not specified
[a]Values not directly reported
[b]Approximate values (taken from graphs)
[c]Normalized by the eluted activity plus activity manipulated during radiopharmacy work
[d]Automated dispensing and injection system (Posyjet)

Table 7.3 Comparison of values of hand skin dose in NM therapy in several published works

Procedure	References	N workers	N measurements	$\langle H_p(0.07)_{max}/A\rangle$ (mSv/GBq) Min	Median	Mean	Max[a]
^{90}Y-Zevalin® preparation	Rimpler et al. [24]	15	1–5	1.2	9.5	11	44
	Rimpler et al. [24][b]	20	1–5	0.3	8.9	39	570
	Rimpler et al. [25]	11	n.s.	2	5.4	–	13(600)
	Geworski et al. [39]	7	n.s.	1.4	–	4.0	8.1
	Cremonesi et al. [40][c]	n.s.	15	0.1	1.5	1.9	28
^{90}Y-Zevalin® administration	Rimpler et al. [24]	19	1–5	1.0	2.9	4.8	12
	Rimpler et al. [24][b]	22	1–5	0.3	3.4	9.0	78
	Rimpler et al. [25]	14	n.s.	0.7	1.0	–	7(27)
	Geworski et al. [39]	8	n.s.	0.4	–	3.3	10.6

For all works, measurements are taken at the maximum (finger tip)
n.s. Not specified
[a]Values in parenthesis are outliers and are not considered in the mean or median calculation
[b]Data including outliers
[c]Values not directly reported

7.3.2.3 Parameters of Influence on Skin Dose to the Hands

Although experimental doses presented high variability, the ORAMED database was sufficient to analyze the main parameters of influence in the measured doses, with appropriate statistical weight [23]. The MC simulation sensitivity study [15] revealed that short source displacements (of up to some few cm), orientation, and volume changes (of up to 3 ml) can increase the maximum dose by a factor from 3 to 5 depending on the source. However, the large range of doses measured for similar techniques means that there is still room for reduction of the largest measured doses.

Shielding was found to be the most important parameter affecting skin dose levels, both for diagnostics and especially for therapy. This result is in agreement with the conclusions of ICRP Publication 106 [22] and with other authors' findings [26–29]. Even though the use of shields slows down the whole procedure, increases the difficulty of visualizing the required volume, and offers less comfort, especially for heavy and thick shields, it provides a protection which mostly cannot be replaced by increasing working speed. The influence of shielding on the dose, estimated in the ORAMED measurement campaign, is shown in Table 7.4. For each procedure, the range, median, mean, and relative standard deviation of the mean of $\langle H_p(0.07)_{max}/A\rangle$ are shown, both for workers using shield and those not using a shield.

For preparation of ^{99m}Tc, it was shown that the influence of the shield on the dose is statistically more significant in the case of the vial than in the case of the

Table 7.4 Range, median, mean and relative standard deviation of mean, $S_{\langle H_p(0.07)_{max}/A\rangle}$ ($k = 1$), of $\langle H_p(0.07)_{max}/A\rangle$ values for all workers in function of the use or lack of use of shield

			$\langle H_p(0.07)_{max}/A\rangle$ (μSv/GBq)					
Procedures	Shield	Number of workers	Range	Median	Mean	±	$S_{\langle H_p(0.07)_{max}/A\rangle}$	Shielding efficiency[a]
^{99m}Tc preparation	Yes	32	30–940	200	310	±	13%	4.3
	No	4	620–2,060	1,290	1,320	±	23%	
^{99m}Tc administration	Yes	24	20–940	90	190	±	27%	1.8
	No	7	140–640	350	340	±	21%	
^{18}F preparation	Yes	18	100–2,930	555	770	±	21%	2.3
	No	12	390–4,430	1,430	1,800	±	22%	
^{18}F administration	Yes	29	110–3,670	630	830	±	18%	5.0
	No	1	4,110–4,110	4,110	4,110	±	0%	
^{90} Y-Zevalin® preparation	Yes	20	0.2–570	7.6	37	±	77%	–
	No	0	–	–	–	±	–	
^{90} Y-Zevalin® administration	Yes	21	0.3–78	2.9	7.4	±	48%	3.1
	No	1	23–23	23	23	±	0%	

Values are rounded

[a]Mean $\langle H_p(0.07)_{max}/A\rangle$ for unshielded data group divided by mean $\langle H_p(0.07)_{max}/A\rangle$ for shielded data group

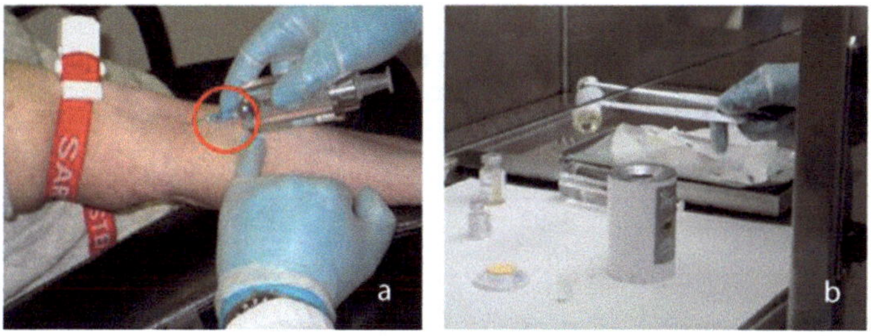

Fig. 7.4 (**a**) Example of bad practices: some fingers are non-properly shielded. (**b**) Example of good practice: tweezers are used to shake the unshielded vial; the procedure is performed in a shielded box

syringe ($p < 0.003$; $p < 0.180$). For the other diagnostic procedures, all workers used shielded vials, and this could not be analyzed. For ^{90}Y-Zevalin, shields are, in general, systematically used (apart from a few outliers), so it is difficult to quantify their influence with the available data. Table 7.4 shows that the use of shields provided a reduction of a factor from 2 to 5 for diagnostic procedures. This reduction factor is much lower than what is calculated using MC calculations for static situations because, in practice, there are always steps during which the shield is not used or situations where part of the hand is not properly shielded. However, MC simulations were found to be very useful to decide which was the adequate shielding for each procedure.

Together with shielding, the literature review shows that the use of automatic devices to avoid worker manipulation is potentially a very efficient mean of dose reduction [18, 29]. Jansson et al. [29], for example, have reported a finger dose reduction of a factor of 5 when using an automatic injection robot with respect to manual injection for ^{18}FDG. However, some works reported some problems associated with the use of automatic devices [26, 29, 30]. Nevertheless, in spite of related problems, most authors agreed on recommending automatic devices for dispensing and injecting, provided that appropriate training was given.

In the ORAMED study, only a weak trend was observed for experience to entail lower doses for diagnostic procedures, but it was not statistically significant. Some studies [31] have shown the positive influence of experience, but it is clear that other issues are more relevant. When analyzing individual cases of high maximum doses, good working habits were found to be more important than experience.

All practices avoiding direct contact whenever possible, enlarging distances to the sources, and speeding up procedures can be considered as good practices. Most bad working habits involved direct source contact. Often staff are not aware that near the bottom of a shielded syringe, the dose rate is very high. One example is given in Fig. 7.4a. Using tweezers is a very effective means of dose reduction when vials or syringes have to be held without a shield (Fig. 7.4b) and also during connecting and separating the syringe to or from needles or butterflies.

Table 7.5 Range, median, and mean values of the ratios between the maximum dose and the dose at the base of the index, base of the ring, and tip of the index fingers calculated for the nondominant (adapted from Carnicer et al. [23] and Sans-Merce et al. [14])

		Maximum dose/dose at other positions			
		Wrist	Base index	Base ring	Index tip
Diagnostics	Range	3–93	2–38	2–60	1–12
	Median	16	4	7	2
	Mean	20	6	10	2
^{90}Y-Zevalin	Range	3–94	2–47	1–87	1–17
	Median	14	7	9	2
	Mean	21	7	15	3

7.3.2.4 Routine Extremity Monitoring

Measurement and simulation results were used for setting up the basis of an appropriate routine monitoring of $H_p(0.07)$ for NM workers. As regards detector technical requirements, Carnicer et al. [16] demonstrate that for ^{99m}Tc measurements, thick standard TLDs (up to 100 mg cm^{-2}) are appropriate, whereas for ^{18}F and ^{90}Y, thin TLDs (up to 10 mg cm^{-2}) are recommended to avoid potential underestimations (up to 50%) because of the beta radiation, as has also been shown in previous studies [32–34].

Dose distribution data were used to find out the best monitoring position. The ratios between the highest dose and the dose at the most common monitoring positions were calculated and are summarized in Table 9.5. It is shown that even with the exclusion of outliers, the distribution of ratios is very wide (from 2 to around 90 for the wrist and from 1 to around 50 for the base of the index).

Mebhah et al. [20] also reported similar ranges, from 5 to 56 for the base and the tip of the middle finger (for diagnostics). This variability responds to the fact that the dose distribution is strongly operator and technique dependent. Thus, taking into consideration the large variations observed, ideally, the best solution for extremity monitoring would be to adapt it to each worker, in other words, to determine, during a trial period, the most appropriate monitoring position for that worker or at least to find out the most exposed hand. If this is not possible, based on the results of the ORAMED project, it is recommended to wear the dosemeter on the index tip of the nondominant hand. However, as there are very few dosimetric systems designed to be situated at this position and since it can cause discomfort, a more practical solution is to wear a ring dosemeter placed on the base of the index finger of the nondominant hand, with the detector facing the palm of the hand. This recommended position is different from other positions proposed in other works such as ICRP 106 [22].

For the recommended monitoring position (base of the index finger), a factor of 6 must be applied to estimate the maximum dose (Table 7.5). Similar correction values were reported by Jankowski et al. [21] and Wrzesien et al. [35]. Other authors [22, 36] published lower ratios, typically from slightly greater than 1 to larger than 4. ICRP 106 [22] recommends for the estimation of $H_p(0.07)$ a dosemeter placed on the base of the middle finger with the element positioned on the palm side. For this position, ICRP recommends a factor of 3 to obtain an estimate of the dose to the tip and of 6 if the dosemeter faces the back of the hand. ORAMED results show that this correction might be too low in many cases. Finally, it should be noted that there is broad agreement that, in nuclear medicine, the ring dosemeter should be preferred to the wrist dosemeter, which underestimates the maximum dose by a factor of 20 [14, 21].

7.4 Recommendations

From the analysis of ORAMED results [14] and other published works on extremity dosimetry in nuclear medicine, nine recommendations are proposed to improve radiation protection of nuclear medicine staff:

1. Extremity monitoring is essential in nuclear medicine. The choice of TLD and TLD position is important for an accurate dose assessment. Thin-layer TLDs (below 10 mg cm^{-2}) are most appropriate when beta-emitters are used.
2. To determine the position for routine monitoring, the most exposed position on the hand for each worker should be found by individual measurements for a short trial period. If for practical reasons, these measurements are not possible, the base of the index finger of the nondominant hand with the sensitive part of the dosemeter placed towards the inside of the hand is the recommended position for routine extremity monitoring in nuclear medicine.
3. To estimate the maximum dose, the reading of the dosemeter worn at the base of the index finger of the nondominant hand should be corrected by a factor of 6.
4. Shielding of vials and syringes is essential. This is a precondition, but not a guarantee for low exposure, since not all parts (e.g., bottom of the syringe) are shielded during use.
5. The minimum acceptable thickness of shielding for a syringe is 2 mm of tungsten for ^{99m}Tc and 5 mm of tungsten for ^{18}F. For ^{90}Y, 10 mm of PMMA completely shields beta radiation, but shielding of 5 mm of tungsten provides better protection, as it cuts down bremsstrahlung radiation.
6. The minimum acceptable shielding required for a vial is 3 mm of lead for ^{99m}Tc and 3 cm of lead for ^{18}F. For ^{90}Y, acceptable shielding is obtained with 10 mm of PMMA with an external layer of a few mm of lead.
7. Any device or tool increasing the distance (e.g., forceps, automatic injector) between the hands/fingers and the source is very effective for dose reduction.
8. Training and education in good practices (e.g., procedure planning, repeating procedures using nonradioactive sources, estimation of doses to be received) are more relevant parameters than the worker's experience level.
9. Working fast is not sufficient; the use of shields or increasing the distance are more effective than working quickly.

Training material and guidelines related to the optimization of radiation protection in nuclear medicine can be downloaded for free from http://www.oramed-fp7.eu/. In addition, the website provides the instructions to receive an easy tool to estimate hand dose distribution for typical nuclear medicine procedures upon acceptance of freeware license agreement.

References

1. Vanhavere F, Carinou E, Donadille L, Ginjaume M, Jankowski J, Rimpler A, Sans Merce M (2008) An overview on extremity dosimetry in medical applications. Radiat Prot Dosim 129(1–3):350–355
2. Donadille L, Carinou E, Ginjaume M, Jankowski J, Rimpler A, Sans-Merce M, Vanhavere F (2008) An overview of the use of extremity dosemeters in some European countries for medical applications. Radiat Prot Dosim 131(1):62–66
3. International Commission on Radiological Protection (1991) Recommendations of the international commission on radiological protection. ICRP publication 60. Ann ICRP 21(1–3)
4. International Commission on Radiological Protection (2007) The 2007 recommendations of the international commission on radiological protection. ICRP publication 103. Ann ICRP 37(2–4)
5. International Commission on Radiation Units and Measurements (1992) Measurements of dose equivalents from external photon and electron radiations, ICRU report 47. International Commission on Radiation Units and Measurements, Bethesda, Maryland
6. International Commission on Radiation Units and Measurements (1993) "Quantities and units in radiation protection dosimetry", ICRU report 51. International Commission on Radiation Units and Measurements, Bethesda, Maryland
7. European Commission (1996) Council directive 96/29/EURATOM of 13 May 1996 laying down basic safety standards for the protection of the health of workers and the general public against the dangers arising from ionizing radiations. Off J Eur Commun 39L:159
8. European Commission (2011) Proposal for a COUNCIL DIRECTIVE laying down basic safety standards for protection against the dangers arising from exposure to ionising radiation. Draft presented under Article 31 Euratom Treaty for the opinion of the European Economic and Social Committee. http://ec.europa.eu/energy/nuclear/radiation_protection/doc/com_2011_0593.pdf
9. Baechler S, Stritt N, Bochud FO (2011) Individual monitoring of internal exposure for nuclear medicine workers in Switzerland. Radiat Prot Dosim 144(1–4):464–467
10. International Commission on Radiation Units and Measurements (1989) Tissue substitutes in radiation dosimetry and measurement. ICRU report 44. International Commission on Radiation Units and Measurements, Bethesda, Maryland
11. Frasch G, Petrova K (2007) Dose trends in occupational radiation exposure in Europe results from the ESOREX project. Radiat Prot Dosim 125(1–4):121–126
12. International Atomic Energy Agency. IAEA training material on radiation protection in PET/CT. Lecture 8 on staff and public doses. http://rpop.iaea.org
13. Whitby M, Martin C (2003) Investigation using an advanced extremity gamma instrumentation system of options for shielding the hand during the preparation and injection of radiopharmaceuticals. J Radiol Prot 23:79–96
14. Sans-Merce M, Ruiz N, Barth I, Carnicer A, Donadille L, Ferrari P, Fulop M, Ginjaume M, Gualdrini G, Krim S, Mariotti F, Ortega X, Rimpler A, Vanhavere F, Baechler S (2011) Recommendations to reduce hand exposure for standard nuclear medicine procedures. Radiat Meas 46(11):1330–1333

15. Ferrari P, Sans-Merce M, Carnicer A, Donadille L, Fulop M, Ginjaume M, Gualdrini G, Mariotti F, Ruiz N (2011) Main results of the Monte Carlo studies carried out for nuclear medicine practices within the ORAMED project. Radiat Meas 46(11):1287–1290
16. Carnicer A, Ginjaume M, Duch MA, Vanhavere F, Sans-Merce M, Baechler S, Barth I, Donadille L, Ferrari P, Fulop M, Gualdrini G, Krim S, Mariotti F, Ortega X, Rimpler A, Ruiz N, Olko P (2011) The use of different types of thermoluminescent dosemeters to measure extremity doses in nuclear medicine. Radiat Meas 46(12):1835–1838
17. Vanhavere F, Berus D, Buls N, Covens P (2006) The use of extremity dosemeters in a hospital environment. Radiat Prot Dosim 118(2):190–195
18. Covens P, Berus D, Vanhavere F, Caveliers V (2010) The introduction of automated dispensing and injection during PET procedures: a step in the optimization of extremity doses and whole-body doses of nuclear medicine staff. Radiat Prot Dosim 140(3):250–258
19. Martin C, Whitby M (2003) Application of ALARP to extremity doses for hospital workers. J Radiol Prot 23:405–421
20. Mebhah D, Djeffal S, Badreddine A, Medjahed M (1993) Extremity dosimetry in nuclear medicine services using thermoluminescent detectors. Radiat Prot Dosim 47(1/4):439–443
21. Jankowski J, Olszewski J, Kluska K (2003) Distribution of equivalent doses to skin of the hands of nuclear medicine personnel. Radiat Prot Dosim 106(2):177–180
22. International Commission on Radiological Protection (2008) Radiation dose to patients from radiopharmaceuticals—addendum 3 to ICRP publication 53. ICRP publication 106. Ann ICRP 38 (1–2) (Annex E: Radiation exposure of hands in radiopharmacies: monitoring of doses and optimisation of protection. International Commission on Radiological Protection)
23. Carnicer A, Sans-Merce M, Baechler S, Barth I, Donadille L, Ferrari P, Fulop M, Ginjaume M, Gualdrini G, Krim S, Mariotti M, Ortega X, Rimpler A, Ruiz N, Vanhavere F (2011) Hand exposure in diagnostic nuclear medicine with ^{18}F and ^{99m}Tc-labelled radiopharmaceuticals—results of the ORAMED project. Radiat Meas 46(11):1277–1282
24. Rimpler A, Barth I, Ferrari P, Baechler S, Carnicer A, Donadille L, Fulop M, Ginjaume M, Mariotti M, Sans-Merce M, Gualdrini G, Krim S, Ortega X, Ruiz N, Vanhavere F (2011) Extremity exposure in nuclear medicine therapy with ^{90}Y-labelled substances -results of the ORAMED project. Radiat Meas 46(11):1283–1286
25. Rimpler A, Barth I, Baum R, Senftleben S, Geworski L (2008) Beta radiation exposure of staff during and after therapies with ^{90}Y-labelled substances. Radiat Prot Dosim 131(1):73–79
26. Montgomery A, Anstee D, Martin C, Hilditch T (1999) Reductions in finger doses for radiopharmaceutical dispensing afforded by a syringe shield in an automatic dose dispenser. Nucl Med Commun 20:189–194
27. Tsopelas C, Collins P, Blefari C (2003) A simple and effective technique to reduce staff exposure during the preparation of radiopharmaceuticals. J Nucl Med Technol 31:37–40
28. Whitby M, Martin C (2005) A multi-centre study of dispensing methods and hand doses in UK hospital radiopharmacies. Nucl Med Commun 26:49–60
29. Jansson B, Olsson S, Olsson B, Jönsson L (2011) Reduction of the absorbed dose to technologists fingers from PET-radionuclides using an automatic injection robot. EANM Annual Congress of the European association of nuclear medicine
30. Allen S, Mackenzie A, Stark G, Inwards G, Lazarus C, Bachelor S (1997) Comparison of radiation safety aspects between robotic and manual systems for the preparation of radiopharmaceuticals. Nucl Med Commun Abstract 18: 295
31. Hilditch T, Elliot A, Anstee D (1990) Fifteen years of radiological protection experience in a regional radiopharmacy. Health Phys 59(1):109–116
32. Carinou E, Donadille L, Ginjaume M, Jankowski J, Rimpler A, Sans-Merce M, Vanhavere F, Denoziere M, Daures J, Bordy J, Itie C, Covens P (2008) Intercomparison on measurements of the quantity personal dose equivalent, $H_p(0.07)$, by extremity ring dosimeters in medical fields. Radiat Meas 43:565–570
33. Ginjaume M, Perez S, Duch M, Ortega X (2008) Comparison of TLD-100 and MCP-Ns for use as an extremity dosemeter for PET nuclear medicine staff. Radiat Meas 43(2–6):607–610

34. Brasik N, Stadtmann H, Kindl P (2007) The right choice: extremity dosemeter for different radiation fields. Radiat Prot Dosim 125(1–4):331–334
35. Wrzesien M, Olszewski J, Jankowski J (2008) Hand exposure to ionising radiation of Nuclear medicine workers. Radiat Prot Dosim 130(3):325–330
36. Covens P, Berus D, Buls N, Clerinx P, Vanhavere F (2007) Personal dose monitoring in hospitals: global assessment, critical applications and future needs. Radiat Prot Dosim 124(3):250–259
37. Tandon P, Venkatesh M, Bhatt B (2007) Extremity dosimetry for radiation workers handling radionuclides in nuclear medicine departments in India. Health Phys 92(2):112–118
38. Leide-Svegborn S (2010) Radiation exposure of patients and personnel from a PET/CT procedure with (18)F-FDG. Radiat Prot Dosim 139(1–3):208–213
39. Geworski L, Zöphel K, Rimpler A, Barth I, Lassmann M, Sandrock D, Zander A, Halm T, Hänscheid H, Hofmann M, Reiners C, Munz D (2006) Radiation exposure in ^{90}Y-Zevalin therapy: results of a prospective multicentre trial. Nuklearmedizin 45(2):82–86
40. Cremonesi M, Ferrari M, Paganelli G, Rossi A, Chinol M, Bartolomei M, Prisco G, Tosi G (2006) Radiation protection in radionuclide therapies with ^{90}Y-conjugates: risks and safety. Eur J Nucl Med Mol Imag 33:1321–1327

Chapter 8
Radiation Doses from Patients to Staff Members, Comforters and Caregivers and to the General Population

Dejan Žontar

8.1 Introduction

After application of a radiopharmaceutical, the nuclear medicine patient contains radioactive material and is a potential source of radiation exposure to various other individuals. Therefore, the risk to critical groups should be assessed and possible precautions taken to limit the exposure of members of the identified critical groups. Individuals may be exposed either in the hospital or outside of it. The individuals at risk in a nuclear medicine department or hospital include three groups:

- Hospital staff (both staff classified as exposed workers and other personnel)
- Other patients
- Persons accompanying or visiting the patient

After the patient is released from the hospital, the critical groups are:

- Comforters and caregivers (relatives and friends who knowingly and voluntarily support and comfort a patient)
- Other household members (both adults and children)
- Visitors
- Work colleagues
- Other members of the public (e.g. fellow travellers, bus and taxi drivers)

The potential risk to other people comes from different modes of exposure:

- External irradiation from photons emitted by the radioactive material in the patient

D. Žontar (✉)
Jožef Stefan Institute, Jamova cesta 39, 1000 Ljubljana, Slovenia

Slovenian Radiation Protection Administration, Ajdovščina 4, 1000 Ljubljana, Slovenia
e-mail: Dejan.Zontar@ijs.si; Dejan.Zontar@gov.si

S. Mattsson and C. Hoeschen (eds.), *Radiation Protection in Nuclear Medicine*,
DOI 10.1007/978-3-642-31167-3_8, © Springer-Verlag Berlin Heidelberg 2013

- Contamination from the contact with radioactive secretions and excretions from the patient (urine, saliva, sweat, faeces, exhaled air, etc.)
- Exposure through environmental pathways (sewerage, discharges to water, cremation of bodies, etc.)

Among the listed modes of exposure, external irradiation is the dominant source of exposure to other individuals [1–3]. The potential risk of radionuclide intake is in general much lower and easier to manage. However, special attention is required in case the patient is breastfeeding. From the point of view of individual doses, environmental pathways are of minor significance. The discharges are dominated by ^{99m}Tc [2] that is of limited importance due to its short physical half-life and by ^{131}I that can be detected in the environment after medical uses, but for which the dose contribution normally is insignificant. The first two modes of exposure will be discussed further in the following sections, while discharges to the environment are addressed in Chap. 10.

Radiopharmaceuticals are applied for diagnostic or therapeutic purposes. In therapy, beta or gamma emitters can be used, while radiopharmaceuticals used for diagnostic imaging are based on gamma-emitting or positron-emitting radionuclides. Positrons are annihilated into two 511-keV photons within a few mm, so from a radiation protection point of view, we are dealing with photons. Pure beta emitters are generally considered to pose small risk of exposure from the patient due to low tissue penetration of electrons. Even if considerable amount of radiation is involved, as is usually the case in therapeutic procedures, the radiation is mostly confined to the patient, his/her excreta and body fluids [2]. Although some research [4, 5] indicates that exposure from beta emitters may be higher than generally assumed, their contribution to the external exposure to other individuals is very small and will not be discussed further in this chapter.

The situation is different in case of gamma- (and positron-) emitting radionuclides. At energies of around 100 keV and above, the photons have high tissue penetration (in case of diagnostic procedures a requirement in the radionuclide selection). For high applied activities, dose rates around the patient may be considerable (e.g. Table 8.1), and irradiation of individuals in the vicinity can be significant. Their exposure is of particular concern for therapeutic procedures where administered activities are high. The most important radionuclide used in medicine that results in the largest doses to the medical staff, comforters and caregivers, relatives and general public is ^{131}I. Other radionuclides used in therapy are mostly beta emitters (e.g. ^{32}P, ^{89}Sr and ^{90}Y) that pose much less risk. In diagnostic procedures, administered activities are lower and the risk of external exposure generally lower. Exceptions are positron emitters (mainly ^{18}F) used in positron emission tomography (PET) that can lead to significant exposure of the staff.

Table 8.1 Influence of distance from the patient on dose rate from patients with thyroid ablation after surgery and follow-up for suspected residual, recurrence or metastases after thyroid cancer treatment at different times after administration (data from [6])

Distance (m)	Day 0	Day 1	Day 2	Day 3	Day 4	Day 7
Mean dose rate per unit activity (μSv/h·MBq) for thyroid ablation patients						
0.1	0.665	0.187	0.088	0.069	0.053	0.016
0.5	0.114	0.049	0.025	0.019	0.014	0.007
1.0	0.046	0.019	0.009	0.007	0.007	0.004
Mean dose rate per unit activity (μSv/h·MBq) for follow-up patients						
0.1	0.746	0.274	0.085	0.030	0.026	0.001
0.5	0.126	0.051	0.017	0.006	0.002	0.0003
1.0	0.046	0.019	0.007	0.003	0.002	0.00014

8.2 External Irradiation

Unless very close to the patient, individuals taking care of the patient are predominantly whole-body exposed in a radiation field that is relatively uniform and gives comparatively low dose rates. The only concern is thus stochastic effects, mainly cancer induction with the lifetime risk for fatal cancer per unit effective dose of approximately 5 %/Sv.

The received dose depends on the radionuclide used, its emission, half-life and biokinetics. Assuming dose rates for ^{131}I as given in Table 8.1 and considering half-lives from Table 8.2 and typical administered activities, it is easy to envisage exposures well above dose limits or dose constraints to the staff, caregivers, family and general public. It is thus important that appropriate precautions and radiation protection measures are applied so that exposure is controlled and minimised. As the main source of exposure is external irradiation, the main controllable parameters available to reduce doses are time, distance and shielding. The key element in minimising the dose is thus knowledge. Hospital personnel that are in regular contact with nuclear medicine patients should be properly trained in radiation protection. For patients and their families, adequate information and instructions should be provided, especially when the patient is released from hospital. The instructions should be based on guidelines for nuclear medicine patients, developed to restrict exposure of family members and general public below dose limits and dose constraints (e.g. IAEA [1], ICRP [2], EC [9]).

8.2.1 Therapeutic Procedures

In therapeutic procedures, radiopharmaceuticals are administered systemically or locally/regionally with an aim to deliver therapeutic doses to the target tissue. According to the UNSCEAR report [10], close to 0.9 million therapeutic nuclear medicine procedures are carried out worldwide each year (compared to 32 million diagnostic procedures yearly). A large majority of those are thyroid therapies,

Table 8.2 Physical half-lives and exposure rate constants for the main radionuclides used in nuclear medicine for diagnostic and therapeutic procedures (Data adapted from [7, 8])

Radionuclide	Physical half-life	Exposure rate constant Γ (μSv·m^2/MBq·h)
^{111}Ag	7.45 days	0.0041
^{198}Au	2.696 days	0.0622
^{11}C	20.4 min	0.148
^{51}Cr	27.704 days	0.0043
^{64}Cu	0.529 days	0.0324
^{67}Cu	2.578 days	0.0157
^{18}F	109.8 min	0.143
^{67}Ga	3.261 days	0.0204
^{68}Ga	68.3 min	0.134
^{123}I	0.55 days	0.0435
^{124}I	4.2 days	0.185
^{125}I	60.14 days	0.0384
^{131}I	8.04 days	0.0595
^{111}In	2.83 days	0.0868
^{13}N	9.96 min	0.148
^{15}O	2.04 min	0.148
^{32}P	14.29 days	Not applicable (β-emitter)
^{82}Rb	76 s	0.159
^{186}Re	3.777 days	0.0054
^{188}Re	0.708 days	0.0070
^{47}Sc	3.351 days	0.0151
^{75}Se	119.8 days	0.0541
^{117m}Sn	13.61 days	0.0400
^{89}Sr	50.5 days	Not applicable (β-emitter)
^{99m}Tc	0.251 days	0.0204
^{201}Tl	3.044 days	0.0121
^{169}Yb	32.01 days	0.0495
^{90}Y	2.67 days	Not applicable (β-emitter)
^{169}Yb	32.01 days	0.0495

specifically hyperthyroidism treatments and treatment of thyroid cancer. A large variety of radiopharmaceuticals is available (Table 8.2) with the most important being those based on ^{131}I.

The level of radiation protection required after administration of radionuclides for therapeutic purposes varies widely depending on the radiopharmaceutical. Those that emit exclusively or predominantly electrons and remain relatively fixed in the body (e.g. ^{89}Sr-chloride, ^{32}P-phosphate, ^{90}Y-Zevalin, ^{153}Sm-EDTPMP and ^{198}Au-colloid) present little risk to other people and in many cases do not require any special precautions. For radionuclides emitting gamma rays, external irradiation of other individuals must be considered. In practice, most issues related to non-occupational exposure from nuclear medicine patients are focused on ^{131}I.

Radioiodine ^{131}I is being applied in thyroid therapies that are the most common therapeutic procedures in nuclear medicine. Typical administered activities are ranging from 100 MBq to 1,000 MBq for hyperthyroidism treatment and from

Table 8.3 Uptake factions and effective half-lives for thyroid ablation (hyperthyroidism treatment) and treatment of thyroid remnants after surgical removal of the thyroid for thyroid cancer (taken from [7])

	Thyroidal component		Extrathyroidal component	
Medical condition	Uptake fraction (F)	Effective half-life (T_{eff}) (days)	Uptake fraction (F)	Effective half-life (T_{eff}) (days)
Hyperthyroidism	0.80	5.2	0.20	0.32
Post-thyroidectomy for thyroid cancer	0.05	7.3	0.95	0.32

4 to 8 GBq for thyroid cancer treatment. Due to the frequent use, high applied activities and high gamma energy (365 keV) ^{131}I is by far the most important radionuclide to consider in radiation protection. The external dose rate from a patient who has got ^{131}I depends on his/her medical condition as this can strongly influence clearance rate from the patient's body and thus the effective half-time (Table 8.1). Clearance is in general much slower for thyrotoxicosis patients than for thyroid cancer patients as their thyroid has been removed and is thus not retaining radioiodine (Table 8.3). At short distances, dose rate will also be influenced by distribution of radioiodine within the patient and that also depends on his/her medical condition.

8.2.1.1 Doses to Hospital Staff

When patients are in the hospital after therapy with radionuclides, the main critical group is hospital staff that may or may not be classified as radiation workers. Although exposure of the staff can be significant, it can generally be managed by appropriate training and well-designed facilities.

The most important critical group is ward nursing staff. They can receive considerable doses from nuclear medicine patients who require hospitalisation after therapeutic procedures. Specially designed shielded isolation rooms with separate toilets reduce the exposure so that it becomes determined mostly by the time spent in the room with the patient. The cumulated dose to nursing staff from a single patient is thus strongly affected by the level of care the patient requires. Although the actual doses vary, they can easily reach a few mSv per year.

8.2.1.2 Doses to Comforters, Caregivers and Family Members

After the patient is released from the hospital, the most important critical groups are caregivers and family members. From a radiation protection point of view, caregivers are people who knowingly and voluntarily accept exposure while helping others undergoing medical diagnosis or treatment, excluding people who do it as part of their occupation. In case of comforters, there is benefit not only for the patient but also for the person helping and comforting the patient; thus, their exposure can be considered medical exposure. ICRP and IAEA thus recommend setting dose constraints instead of rigid dose limits [1, 2]. Although there are some variations, the recommended values are in the range of a few mSv per episode (IAEA: 5 mSv/episode; EC: 1 mSv for children, 3 mSv for adults under 60 years and 15 mSv for adults above 60 years [9]).

Among family members, most attention should be given to exposure of children, particularly infants young enough to be held in close contact for prolonged periods of time. A rough estimation of the absorbed dose from close contact can be obtained from measured surface dose rate before release from the hospital and effective exposure times such as those given in [3]. Various studies (summarised in [1–3]) show that doses to infants can be kept below 1 mSv as long as parents comply with instructions. However, if safety measures are not observed, dose limit can be significantly exceeded.

Another group that can largely exceed dose limits, if safety measures are not followed, are partners of patients administered with therapeutic activities of radionuclides. Although behaviour during daytime is important, one of the main factors affecting their exposure appears to be related to their sleeping arrangements. Various studies (summarised in [3]) report that doses from night period were a few times larger than those from daytime period if partners were sharing patient's bed. The results indicate that after ^{131}I thyrotoxicosis therapy, dose to the partner may exceed 5 mSv if there are no restrictions to sleeping arrangements after patient's return home. Although exposure strongly depends on a number of parameters that vary significantly by country and families within a country, IAEA and ICRP published some general recommendations on the duration of separate sleeping arrangements [1, 2].

Actual measurements show that in practice exposure of family members and carers rarely exceeds a few mSv and stays within recommended dose limits and constraints as long as instructions are being followed [1–3]. IAEA [1] thus recommends that restrictions should focus on special groups such as pregnant women, infants and children.

8.2.1.3 Doses to the General Public

Outside of patient's home, critical groups include members of the general public, the most important being work colleagues and people encountered in public places, particularly during transport. Dose limit recommended by the ICRP and IAEA for members of the public is 1 mSv/year. In some countries (e.g. EC [9]), additional dose constraints that may be as low as one third of the dose limit are applied to individual planned activities involving members of the public.

Travellers sitting close to a nuclear medicine patient on his/her travel from the hospital will be exposed to external irradiation. The received dose will depend on the retained activity in the patient, travelling time and distance between the patient and the fellow traveller. When private transport is used and the driver is sitting at a distance of about 1 m from the patient, restrictions regarding travel time are normally only required for thyroid cancer patients within 48 h after radionuclide administration [2]. In case of patients using public transport, doses to close-sitting fellow travellers can be significant, and time restrictions based on administered activity and type of the procedure have been recommended, for example, [1, 2]. In general, it is however unlikely for patients to present a risk to other passengers for travel times within a few hours [1].

Exposure of co-workers will depend on administered activity, time between the procedure and return to work and distance from the patient. It is somewhat easier to manage as work colleagues can be informed of the patient's condition and safety measures. Total dose can be estimated from dose rate measurements and work pattern. Recommended number of days off work after ^{131}I therapy based on administered activity, type of work, etc. is given in [9].

8.2.2 Diagnostic Procedures

A wide variety of radiopharmaceuticals are administered for diagnostic purposes (Table 8.2), and according to UNSCEAR [10], more than 32 million diagnostic nuclear medicine procedures are carried out worldwide each year. A large majority of procedures use radiopharmaceuticals labelled with ^{99m}Tc and are performed with gamma cameras or SPECT systems. Among those, the most frequent are bone scintigraphy and myocardial, lung and brain perfusions [10]. The other important group of radiopharmaceuticals is based on positron-emitting radionuclides and intended for PET imaging. Although a number of radionuclides suitable for PET imaging is available (e.g. ^{18}F, ^{11}C, ^{13}N, ^{15}O, ^{64}Cu, ^{68}Ga, ^{82}Rb and ^{124}I), the majority of radiopharmaceuticals are labelled with ^{18}F.

From the radiation protection point of view, one should distinguish between radiopharmaceuticals based on gamma-emitting radionuclides and those based on positron-emitting radionuclides used for PET. Due to higher penetration of the annihilation photons (HVL in lead is about 4 mm for 511-keV photons and 0.3 mm for 140-keV photons) in PET, not only radiation safety in the vicinity of the patient but also in adjacent rooms should be considered. Shielding in PET facilities is thus a much more challenging issue, and shielding designed for gamma cameras and SPECT is generally not adequate. For PET, shielded areas should include all regions where patients spend considerable time such as uptake rooms, imaging room and post-imaging waiting area. Shielding may be required not only for walls but also for ceiling and floors. As shielding is not the focus of this chapter, the reader is advised to consult Chap. 10 or other sources, e.g. [8, 11, 12], for more information related to design and shielding of PET facilities.

When evaluating dose rate from a patient, the effective shielding due to absorption in the patient's body should also be taken into account. In case of positron emitters, the body absorption factor is often set to 0.36 [8, 11]. Based on a survey of dose rate measurements, a dose rate from PET patients immediately after administration of ^{18}F was estimated to 0.092 μSv/MBq at 1 m [8]. During the first hour after injection, the dose rate decreases by approximately 30 % due to voiding (in case of ^{18}F-FDG, approximately 15–20 % of administered activity is excreted within the first 2 h [8]) and physical decay.

8.2.2.1 Doses to Hospital Staff

The main critical group for diagnostic nuclear medicine procedures are imaging technologists. Studies (summarised in [3]) show that a large contribution to their total dose is due to external irradiation from patients. Even in departments where technologists dispense and inject radiopharmaceuticals as well as operate the imaging equipment, dose from the patient is larger than dose from the syringe while dispensing and injecting. Imaging technologists are exposed to external irradiation while performing various tasks near a post-injection patient during the uptake period, while escorting the patient to and from the scanner and while positioning him at the scanner or on the scanner bed.

For ^{99m}Tc investigations, an average dose to imaging technologists is of the order of nSv/MBq per procedure [3]. This dose depends on a number of factors such as administered activity, time between injection and scan, patient size, individual technique of a technologist, shielding and last but not least patient cooperation. While an annual dose will depend on additional factors including workload, typical values for whole-body dose of about 0.3–0.4 mSv are reported by IAEA [11].

Imaging personnel working on PET generally receive larger annual doses than those working with gamma cameras and SPECT. They are becoming one of the medical professionals with the highest exposure along with the radiological and cardiological interventionists. During patient positioning, the dose rate 1 m from a patient administered 555 MBq of ^{18}F (a typical ^{18}F-FDG study) is around 30 μSv/h [8]. Reported doses to PET technologists are around 5 μSv per procedure and around 8 mSv annually [8, 11, 12]. Due to large influence of patient's ability to cooperate, this can be even one of the factors taken into account when deciding about referral for a PET investigation.

Doses to the ward nursing staff from diagnostic procedures are significantly lower than from therapeutic patients [3]. As in case of therapeutic procedures, it is strongly influenced by required level of care. The dose contribution from ambulant patients is of the order of μSv/day, while for totally helpless patients who require hospitalisation and high level of care, it can reach around 100 μSv/day.

8.2.2.2 Doses to Caregivers and Family

Special measures to reduce exposure of relatives and general public are usually not necessary after diagnostic procedures with gamma-emitting radionuclides used in planar imaging and SPECT. Exceptions are breastfeeding patients and patients who were subjected to whole-body scans with ^{131}I for recurrent thyroid cancer. Attention should however be paid to young infants who require extended periods of close contact. Dose estimates for infants indicate that their exposure should remain under 1 mSv for most diagnostic procedures assuming typical administered activities although in some situations periods of close contact should be restricted [3]. In a reverse situation, dose to parents from an infant subjected to a nuclear medicine diagnostic procedure appears to be even less likely to exceed 1 mSv [3].

The main radionuclides used in PET investigations have short physical half-lives (under 2 h, Table 8.1), so dose rate from a patient diminishes rapidly with time. Exposure to caregivers, family and general public can thus only be considerable within a few hours after administration. Thus, a caregiver accompanying patient in an uptake room after radiopharmaceutical administration could receive radiation dose of the order of 0.1 mSv.

A large collection of dose estimates from nuclear medicine patients for various real-life situations, based on questions from concerned individuals, can be found on "Ask the Experts" pages of the Health Physics Society [13].

8.3 Dose Assessment Methods

The basis for development of general recommendations on safety measures is assessment of doses to critical groups. Two different approaches are used: measurement of cumulated dose received by an individual and calculation from measured dose rates at different distances and from duration of exposure at those distances. In the first approach, the dose to an individual is determined by him/her (e.g. caregiver or family member) wearing an integrating dosimeter (e.g. TLD) during the period of exposure (e.g. from release of the patient from a hospital till the "total" decay of the radiopharmaceutical). The advantage of this method is that it integrates the actual dose rates, taking into account decrease with time and with distance from the patient. The precision of this approach is limited by how closely the selected individual complies with the instruction for wearing dosimeter (such as when and where on the body to wear it) and by non-uniform energy response of the detector. Sensitivity of the dosimeter may be another limiting factor when the aim is to assess dose from an individual patient.

In the second approach, measurements of time-averaged dose rates at different distances and typical times spent at those distances are used as an input data for calculation of a cumulative dose to an individual from a critical group. The advantage of this approach is that it does not require cooperation of the individuals from the critical group for which the assessment is being made. Besides, the same dose rate measurements can be used to assess doses to any critical group. The drawback is that accuracy strongly depends on validity of assumptions about the behaviour of individuals from the selected group, influencing times and distances used as an input to the model. It also depends on the validity of a model describing decrease of dose rate with time and relies on interpolation or extrapolation of dose rate data for distances where measurements have not been made.

The approaches described above are tools for dose assessment for typical groups from an average patient. The results are used as a basis for developing recommendations, particularly on release of patients after therapy with radionuclides. While measurement of the cumulated dose could in principle be used to determine dose from a particular patient to any given individual, it is not practicable on regular basis. On the other hand, estimation of dose from an individual patient based on calculations is practicable. A realistic dose calculation combined with dose limits and constraints as a basis for decision on patient's release is recommended by the main authorities in radiation protection such as IAEA [1], ICRP [2] and US NRC [7]. An introduction into external dose calculations is given in the next section.

8.4 Calculation of External Dose

The basic models, used for calculation of dose to exposed individuals due to external irradiation from a patient after administration of a radionuclide, are rather simple. However, input data are based on numerous assumptions, so dose calculations are far from straightforward and subject to many uncertainties. This may lead to large variations in calculated doses depending on the model and the assumptions used.

A number of parameters have to be considered when calculating dose due to external irradiation from a patient. The most important among them are:

- Properties of the radionuclide (decay mode and energy, physical half-life)
- Applied and/or retained activity
- Metabolism of the radiopharmaceutical (physiological half-life, distribution of the radioisotope in the patient)
- Distance
- Time
- Attenuation

- For the purpose of dose calculations, properties of a radionuclide are usually comprised into two constants:
- Exposure rate constant (specific gamma ray constant for a point source) Γ represents dose rate 1 m from a point-like source of a given radionuclide with activity of 1 MBq.
- Physical half-life $T_{1/2}$ describes the rate of radioactive decay $A = A_0 2^{-t/T_{1/2}}$.

Exposure rate constants and half-lives of the main radionuclides used in nuclear medicine are listed in Table 8.1.

The basic model for dose rate calculations assumes a point-like source with initial activity A_0 and exposure rate constant Γ. Another assumption is that time dependence of activity in the patient is governed only by the radioactive decay with a physical half-life $T_{1/2}$. In such a simplified case, the dose rate D at distance r from the source at time t after administration is described by the equation

$$\dot{D}(t) = \Gamma A_0 2^{-t/T_{1/2}} \frac{1}{r^2}. \tag{8.1}$$

For purposes of radiation protection, one is usually less concerned with the instantaneous dose rate from the patient and more often with the total dose received by the staff and the public. For this purpose, (8.1) can be converted into integrated form

$$D(t) = 1.44\Gamma T_{1/2} A_0 (1 - 2^{-t/T_{1/2}}) \frac{1}{r^2}, \tag{8.2}$$

where $D(t)$ represents a total dose at a distance r integrated from time 0 to t. The newly appearing factor 1.44 is a result of integration of the exponential. It should be noted that half-life is often given in days, while the gamma constant is usually given in μSv· m^2/MBq· h. To avoid conversion of the half-life from days to hours, the integration constant 1.44 is often combined with the conversion factor 24 h/day, so the constant 1.44 is replaced by 34.6. Although practical, such approach introduces mismatched units (i.e. days for $T_{1/2}$ and hours for Γ) and will be avoided in this chapter.

Another commonly used approximation is to integrate from time 0 to infinity ($t \to \infty$) and to describe the actual time behaviour by an occupancy factor E, in which case (8.2) turns into

$$D(\infty) = 1.44\Gamma T_{1/2} A_0 E \frac{1}{r^2}. \tag{8.3}$$

The occupancy factor E represents the fraction of the total dose received by an individual as a consequence of the actual time spent in the vicinity of the source (patient).

Although very basic, the above equation is often used for dose estimation. Due to many simplifications and conservative assumptions, it generally gives a very conservative estimation of the received dose. When used for dose reconstruction, it can lead to significant overestimation of the received dose. When used for dose planning, it can lead to overly restrictive radiation protection measures increasing cost and/or reducing comfort of the patient. In the following sections, different methods for estimation of the main factors affecting dose calculation will be briefly described.

8.4.1 Duration of Exposure and Occupancy Factor

Two of the factors affecting total dose received by an individual are duration of exposure and time between administration and exposure. In case of dose reconstruction (estimation of radiation dose received by an individual in the past as a result of particular exposure situations of concern) these data may be at least approximately known. Therefore, dose rate integration over the appropriate time intervals (or its estimate from interval length and approximate average dose rate within the interval) can be performed. When estimating future exposure, for example, when making a decision about the release of a patient from a hospital, this information is not available and can only be assumed.

Calculation of retained activity at which patients may be released from a hospital, published by US Nuclear Regulatory Commission [7], is based on the following assumptions about occupancy factors at distance 1 m from a patient:

- $E = 0.25$ for radionuclides with physical half-life >1 day.
- $E = 1$ for radionuclides with physical half-life shorter than or equal to 1 day. The guide argues that long-term averaging of behaviour cannot be assumed due to shorter half-life. The total dose is most affected by the time an individual spends close to the patient immediately after his release, and in case of short-lived radionuclides, this can amount to a large fraction of the total dose. Thus, for short half-lives, occupancy factors from 0.75 to 1 are considered appropriate.

This approach, combined with the fact that no biological elimination of the radionuclide is taken into account, is conservative in most normal situations. It can be used for conservative estimation quite generally with little consideration of specific details. The conservativeness in setting the above occupancy factors can be seen by comparison with the effective exposure times as given in [3].

When more realistic dose estimation is preferable (and practicable), the calculation should be based on the patient-specific information. When determining the occupancy factor for the purpose of calculations of maximal retained activity at patient release, various considerations should be taken into account. Some of them are:

- Does the patient live alone, with adult members of the household or with children (especially in case of small children, it is less likely that prudent distance will be maintained)?
- Is separate bedroom available for the patient or will he/she share a bed?
- Expected number and duration of visits by family or friends.
- Duration and means of transport (personal transport, public transport, etc.).

8.4.2 Distance and Attenuation

In dose calculations, activity distribution within a patient is often assumed to be an unattenuated point source. For hyperthyroid patients or thyroid cancer patients with localised metastases, this assumption is close to reality, and inverse square law is generally considered a good approximation of dose rate dependence even at close distances. Measurements of dose rates from patients with thyroid cancer treatment were performed by Barrington et al. [6]. Table 8.2 shows that even for those patients, inverse square law is only an approximation and that in practice, dose rate dependence on distance is somewhat weaker.

In many nuclear medicine procedures, such as palliative treatment of bone metastases, radioimmunotherapy or PET imaging with ^{18}F-FDG, activity is widely distributed over the patient. In radioimmunotherapy, most of the activity is located within the torso, while in radionuclide therapy involving bone disease and in imaging with ^{18}F-FDG, it may be distributed along the entire patient. When activity is distributed over larger volumes, dose rate is decreasing with distance at a slower rate than for a point source. When extrapolating from measurements at longer distances, assuming a point-source distribution can lead to a significant overestimation of dose rates closer to the patient. A linear-source model with attenuation correction can be used to more accurately reflect activity distribution for some procedures. In a line-source model as proposed by Siegel et al. [14] the falloff of the dose with distance depends on both distance and length of the source. In such geometry, the $1/r^2$ term in equations (8.1–8.3) can be replaced by an arctangent term:

$$\frac{1}{r^2} \rightarrow \frac{2\arctan(l/2r)}{lr}, \tag{8.4}$$

where l represents the length of a line source. A comparison of point-source model with line-source model is given in [14], including a table of conversion factors for different source lengths and distances from the source. The factors range from 0.997 for position 100 cm from a 20 cm long source to 0.167 for a point 10 cm away from a 174 cm long source. The listed conversion factors give an indication of the level of overestimation that can result from using a point-source model indiscriminately. For longer distances, both measurement and Monte Carlo simulations have shown that inverse square law represents an adequate description of the dose rate variation with distance [3]. In case of the line-source model described in [14], results obtained by inverse square law approximation and the line-source model are within 10 % for distances longer or approximately equal to the source length. Although more accurate results may be obtained by more complex models, the advantages of the line-source model are argued to be improved accuracy as compared to a point-source model while maintaining ease of implementation and conservative approach.

Equations for dose rate or cumulated dose discussed above do not account for attenuation and scatter in the patient, although they can significantly reduce the dose rate around a patient. The attenuation may be accounted for by multiplying the exposure rate constant Γ by an appropriate factor (e.g. 0.6 for ^{131}I [14] and 0.64 for ^{18}F [11]) or through theoretical calculations. The third, most reliable method is by measurement of the dose rate at a given distance, from which the ΓA_0 term that includes attenuation correction can be obtained.

None of the models discussed above take into account directional dependence of dose rate. It can result from inhomogeneous activity distribution within the patient and from differences in absorption in different directions. Measurements showed significant variations in dose rate in anterior, posterior and lateral directions [3].

8.4.3 Time Dependence

After application of a radiopharmaceutical, concentration of a radioisotope in any given organ changes with time. Time dependence is a consequence of radioactive decay of the radioisotope and of physiological processes that move the radiopharmaceutical around in the body. For a given organ, the time variation of the concentration of the radiopharmaceutical can be modelled by an exponential function (or more often a set of exponentials) that describe the uptake (increase of concentration in an organ) or elimination (decrease of concentration in an organ).

When calculating doses due to radiation from the patient, both processes should be taken into account. Time dependence is often described by effective half-life (T_{eff}) that combines contributions of various processes. In terms of decay constants, the effective decay constant (λ_{eff}) can be expressed as a simple sum of physical decay constant ($\lambda_{1/2}$) and biological decay constant (λ_{b}):

$$\lambda_{\text{eff}} = \lambda_{1/2} + \lambda_{\text{b}}. \tag{8.5}$$

In terms of physical half-life of the radionuclide ($T_{1/2}$) and the biological half-life for physiologic elimination (T_{b}), the effective half-life ($T_{1/2}$) can thus be expressed as

$$T_{\text{eff}} = \frac{T_{1/2} T_{\text{b}}}{T_{1/2} + T_{\text{b}}}. \tag{8.6}$$

A simple single-exponential model is generally a good description of the physical decay as radioisotopes with long decay chains are usually avoided in nuclear medicine due to extra exposure of the patient from the daughter radionuclides. On the other hand, physiological processes usually require models with more than one exponential for an accurate description.

8.4.3.1 Multi-component Calculations

Multi-component calculations will be exemplified by a model for ^{131}I because this is the radioisotope for which more precise calculations are most likely to be needed. Behaviour of the ^{131}I can be modelled using two components:

- Iodine outside of the thyroid (extrathyroidal iodine)
- Iodine in the thyroid after uptake (thyroidal iodine)

An effective lifetime can be calculated for each component using the same physical half-life but separate biological half-times. A two-component model is thus described by two exponentials with appropriate half-lives and weights (uptake fractions). Values for uptake fractions and effective half-lives for thyroidal and extrathyroidal component for two most common forms of nuclear medicine therapy, thyroid ablation (hyperthyroidism therapy) and treatment of thyroid remnants after surgical removal of the thyroid (thyroid cancer therapy) as suggested by US Nuclear Regulatory Commission, are given in Table 8.3. A large difference can be observed between those two types of procedures because in the latter, thyroid is removed and is not retaining radioiodine. As a consequence, from cancer patients, a vast majority of radioiodine is removed via biological pathways (urine) in the first 2 days after treatment. This is modelled by a large uptake fraction (95 %) and short effective half-life (0.32 days) of the extrathyroidal component. It is obvious that neglecting the physiologic processes and performing the calculation based only on physical half-life (8 days) will lead to gross overestimation of the dose.

The described two-component model does not take into account the time for the iodine to be absorbed from the stomach to the blood and the hold-up of the iodine in the bladder. When calculating exposure of other individuals from external radiation from a patient treated by ^{131}I, neglecting those contributions can lead to an underestimation of the received dose. A possible solution, recommended by US Nuclear Regulatory Commission, is to conservatively assume that during the first 8 h after administration, about 80 % of the administered activity is removed from the body with an effective half-life determined only by the physical half-life ($T_{eff} = T_{1/2}$). Thus, an equation to calculate the total dose, received by external radiation from a patient treated with ^{131}I, has three components:

$$D(\infty) = 1.44\Gamma A_0 E \frac{1}{r^2}\left\{E_1 T_{1/2}(0.8)\left(1-2^{-8h/T_{1/2}}\right)+E_2 2^{-8h/T_{1/2}}\left(F_1 T_{1eff}+F_2 T_{2eff}\right)\right\}. \quad (8.7)$$

In the above equation, the first component gives the dose received in the first 8 h after administration using occupancy factor E_1, weight 0.8 (80 %) and physical half-life $T_{1/2}$. The second and the third components represent dose received from 8 h after administration to total decay. As the starting activity for those two components, activity at 8 h after administration calculated using only radioactive decay is assumed. For each component, appropriate uptake factor (F_1 or F_2) and effective half-time (T_{1eff} or T_{2eff}) are used, while the occupancy factor E_2 is the same.

8.4.3.2 Comparison with Measurements

In practice, it is more accurate to determine time dependence of dose rate from a series of measurements than to calculate it from a model based on one or more effective half-lives. One of the reasons is that the computational models, even if they include multiple components, are based on whole-body retention of the radioactivity and do not take into account anatomical redistribution of radioactivity with time. Such redistributions may influence dose rates at different distances and lead to different time dependence of dose rates at those distances. The variation of dose rate with distance is influenced by physiological processes in the patient (and thus by administered radiopharmaceutical and patient's pathology) and by the size of the patient [3].

8.4.4 Influence of Models and Assumptions on Calculated Dose

Various authors [4, 8, 15] assessed accuracy of dose calculation models by comparing calculated doses with actual measurements. Consistent with the limitations described above, the reports show that a simple point-source model significantly overestimates dose rate at short distances, while calculations based on more complex models gave good agreement with measurements.

It is however not only the complexity of the model used for dose calculation that affects accuracy of the calculated dose. As the input data are generally at least to some extent based on assumptions, those assumptions can be a source of significant errors as well. An educative example is reconstruction of absorbed dose for a family member of a ^{131}I therapy patient as studied and commented in [16].

8.5 Internal Exposure

Although the major path of exposure for all critical groups is external irradiation, contamination from the patient is also a potential source of exposure of medical staff, caregivers and relatives. Some radionuclides, ^{131}I among them, have multiple excretion routes from the body before they decay. Those radiopharmaceuticals can expose other people and the environment to irradiation and contamination unless precautions are taken.

Radioactive iodine is excreted primarily in the urine with smaller amounts in saliva, sweat, faeces and small amount in exhaled air. Clearance rates vary for different pathways, radiopharmaceuticals and between patients. For some treatments, a significant portion of the administered activity is discharged to drains, up to approximately 90 % in case of thyroid carcinoma (^{131}I), bone metastases (^{89}Sr chloride) and phaeochromocytoma (^{131}I MIBG). In thyroid carcinoma treatments, about 85 % of the applied radioiodine is discharged in the first 5 days, with over 50 % discharged in the first 24 h [2]. Under normal conditions, internal contamination through direct contact with excreted ^{131}I is small, and resulting doses are about two orders of magnitude lower than from external irradiation. In case of incontinent patients and patients that are not able to urinate without spilling urine, this pathway may however be of concern, and adequate measures should be applied.

Another possible pathway of internal contamination is inhalation of ^{131}I aerosols exhaled by the patient. Although this pathway is not always recognised, it can give substantial dose not only to the individuals close to the patient but to all persons present in the same room. Studies show that in radioiodine therapy up to 1‰ of the administered activity can be released into the air in the therapy room [2]. Activities in exhaled breath can reach a few 100 Bq/l and in the room air up to few tenths of

Bq/l. Thyroid doses to the hospital staff around few mGy/year were reported for well ventilated rooms.

8.5.1 Contamination

Radioactive excretions are most likely to be expressed as contamination of eating and drinking utensils, toilets and bed linen. At release from the hospital, appropriate instructions should be given to the patient depending on the type of procedure, time from the procedure (particularly whether patient is released immediately after procedure or after a few days of hospitalisation) and social circumstances in which the patient lives. In case of oral administration, instructions should generally include sickness or other circumstances that may result in vomiting.

Contamination from a thyroid cancer patient is the greatest at approximately 1 day after administration and is higher than from patients undergoing therapy for hyperthyroidism. Removable activity from hands is strongly affected by how frequently hands are washed but can reach a few 100 Bq/cm^2. Removable activity from surfaces a patient touched is also very variable but can reach similar levels. For male patients, contamination can be much higher on toilet rims (order of kBq/cm^2). Isolation rooms in hospitals thus often need to be decontaminated before they can be used by other patients. Hospital staff and particularly nurses should be properly trained and act with caution when dealing with vomit and coughing or sneezing patients.

The maximal likely effective dose from internal contamination can be roughly estimated from activity administered to the patient (A_0), assumed intake fraction (F_U) and dose conversion coefficient that converts internal activity into effective dose (f_{DC}) using equation

$$D_i = A_0 F_U f_{DC}. \tag{8.8}$$

In this approach, the main source of uncertainty lies in estimation of the intake factor. In general, it can be assumed that the intake of individuals exposed to patients administered ^{131}I is of the order of 10^{-6} [1, 2, 7]. As a conservative approach, US NRC [7] recommends to assume an uptake fraction of 10^{-5}. More reliable approach than use of the intake fractions is measurement of activity in body fluids or in people other than the patient. Dose conversion coefficients for various radionuclides and age groups can be found in tables. For inhalation and ingestion of ^{131}I by adults, they equal 7.4×10^{-9} Sv/Bq and 2.2×10^{-8} Sv/Bq, respectively [2, 9]. The maximal estimated internal dose from a patient who was administered 1 GBq of ^{131}I would thus be of the order of 0.1 mSv. An overview of activities for different paths of contamination is available in [2]. Detailed information about risks and recommended precautions for a variety of nuclear medicine procedures are provided by IAEA [1].

Even under conservative assumptions, the estimated internal dose for adults is usually well below that from external irradiation and can in most situations be neglected in calculations of the total dose. This may not be the case for cancer treatments with ^{131}I where a large part of applied activity is excreted (see Table 8.3) reducing the external dose and potentially increasing internal contamination. Risk of thyroid cancer induction due to radioiodine contamination for adults is low. Thyroid glands of infants and foetus are much more sensitive to induction of thyroid cancer than in adults, and doses from internal contamination are not necessarily negligible compared to doses from external irradiation. In normal conditions, protection measures should thus be focused towards avoiding contamination of pregnant women and children [1, 2].

8.5.2 Breastfeeding

After administration of a radiopharmaceutical to a breastfeeding woman, the radionuclide will be secreted in her milk. If fed to an infant, such milk will cause internal contamination and consequently dose to the infant. The problem of breastfeeding is addressed in detail in Chap. 10.

8.6 Special Circumstances

8.6.1 Borders and Airports

Nuclear medicine patients may have measurable emissions of gamma radiation for some time after administration, for weeks in case of ^{131}I therapy. Even if dose rates are well below levels of concern for health, the radiation may be detected by radiation monitors at airports, borders and other places where they may be used for security purposes. In such cases, patients should be issued documents confirming that they had a medical treatment/investigation using radioactive material. It should include information about administered activity and radionuclide. While staff operating security detectors should be trained in how to deal with nuclear medicine patients, such documents may not be recognised by poorly trained security personnel. Patients should be advised to carry the document when travelling. It may also be preferable to avoid air travel and border crossing unless they are willing to accept some inconvenience.

8.6.2 Immediate Medical Interventions

Nuclear medicine patients who require emergency surgery, suffer from a heart attack or require other life-saving procedures should be treated in the same manner as any other patient even if residual activity is high. Such situations can lead to significant amount of contamination and to significant dose to personnel from external irradiation. Procedures may require involvement of personnel with no training in radiation protection and may be performed outside the nuclear medicine ward. Nuclear medicine staff should thus provide guidance to involved personnel. Radiation protection officer should be involved as soon as possible, and advice should be sought from medical physicist if available. The number of involved personnel should be limited to those required, and involved personnel should be indentified in case dose reconstruction is required. Proper care should be taken of contaminated material (body fluids, medical waste, laundry, etc.). Decontamination of the rooms, instruments and/or equipment may be necessary.

If a patient requires emergency care after he was released from the hospital, the same procedure should be followed as if he was hospitalised. It is the duty of the patient or accompanying persons to immediately inform medical personnel that the patient had nuclear medicine therapy. Compared to hospitalised patient, available information is usually more limited and possibly less reliable, making decisions and eventual dose reconstruction more difficult.

Doses to the staff resulting from medical interventions can vary greatly depending on administered activity and radionuclide, time from administration, medical condition, treatment and other factors but could reach around 100 μSv for the more exposed individuals.

8.6.3 Post-mortem Examinations and Burial

In case of death of a patient who recently had a therapy with radionuclides, special considerations may need to be given to the treatment of the corpse. A qualified radiation protection expert should be consulted to determine appropriate precautions, and keep doses as low as reasonably achievable. When a patient dies after release from a nuclear medicine department, details of his treatment may not be readily available. Whenever possible, such information should be obtained from the hospital that treated the patient. As use of radionuclides in palliative treatment is increasing, the number of such cases can also be expected to increase.

Radioactive bodies should be clearly identified as potential hazard, and body bags may need to be used to contain leakage of radioactive substances. Depending on surrounding dose rate, the body may need to be kept in a radiation controlled area, and close contact with the deceased should be minimised. Controls may also be required on subsequent stages of disposal, and involved persons should be informed of the circumstances and advised of any necessary precautions. Despite this, wishes of next of kin should be considered, and dignity should be maintained whenever possible.

The main areas of concern are autopsy examinations, embalming and burial or cremation of the corpse. Restrictions for some of those situations are summarised in [2].

Autopsy examinations, particularly on patients that recently received ^{131}I for thyroid cancer therapy, may lead to significant exposure of pathologists. Necessary precautions should be determined in cooperation with a radiation protection expert. Prior to the autopsy, consideration should be given to the removal of contamination and disposal of resulting radioactive waste. As doses to pathologists may reach hundreds of μSv to whole body and several mSv to hands, it may be a good practice to divide the work among more pathologists in order to keep individual doses low. In some cases, it may also be advisable to postpone the examination for a short time period. Similar restrictions may be required for embalming of the corpse.

Depending on national regulations, burial or cremation may need to be delayed, or radiation protection measures may be required in case of high retained activities. In cremated remains, a part of retained activity will be present, and scattering of ashes may be an issue for long-lived radionuclides (^{89}Sr). A review of some national requirements is available in [1, 2].

References

1. International Atomic Energy Agency (2009) Release of patients after radionuclide therapy. Safety report series no. 63
2. The International Commission on Radiation Protection (2004) Release of patients after therapy with unsealed radionuclides. ICRP publication 94. Ann ICRP
3. Mountford PJ, O'Doherty MJ (1999) Exposure of critical groups to nuclear medicine patients. Appl Radiat Isot 50:89–111
4. Meo SL, Cicoria G, Montini G, Bergamini C, Campanella F, Pancaldi D, Panebianco AS, Rubello D, Marengo M (2008) Radiation emission dose from patients administered ^{90}Y-labelled radiopharmaceuticals: comparison of experimental measurements versus Monte Carlo simulation. Nucl Med Commun 29(12):1100–1105
5. Rimpler A, Barth I, Sentfleben S, Baum RP, Geworski L (2008) Beta radiation exposure of staff during and after therapies with Y-90 labelled substances. EURADOS Annual meeting 2008. http://www.eurados.org/meeting2008/Workshop/03-Rimpler.pdf
6. Barrington SF, Kettle AG, O'Doherty MJ, Wells CP, Sommer EJR, Coakeley AJ (1996) Radiation dose rates from patients receiving iodine-131I therapy for carcinoma of the thyroid. Eur J Nucl Med 23:123–130
7. Office of Federal and State Materials and Environmental Management Programs, US Nuclear Regulatory Commission (2008) Consolidated guidance about materials licenses: program-specific guidance about medical use licenses. NUREG-1556, vol 9, Rev. 2
8. Madsen MT, Anderson JA, Halama JR, Kleck J, Simpkin DJ, Votaw JR, Wendt RE III, Wiliams LE, Yester MV (2006) AAPM task group 108: PET and PET/CT shielding requirements. Med Phys 33(1):4–15
9. European Commission (1998) Radiation protection following iodine-131 therapy. Radiation protection 97
10. United Nations Scientific Committee on the Effects of Atomic Radiation, UNSCEAR 2008 report vol 1 Sources of ionising radiation
11. International Atomic Energy Agency (2008) Radiation protection in newer medical imaging techniques: PET/CT. Safety report series no. 58
12. Hippelainen E, Nikkinen P, Ihalainen T, Uusi-Simola J, Savolainen S (2008) Personal radiation doses in PET/CT facility: measurements vs. calculation. Radiat Prot Dosim 132(1):57–63
13. Health Physics Society. Ask the experts web page. http://www.hps.org/
14. Siegel JA, Marcus CS, Sparks RB (2002) Calculating the absorbed dose from radioactive patients: the line-source versus point-source model. J Nucl Med 43:1241–1244
15. Monfared AS, Amiri M (2010) Physical dosimetry and mathematical dose calculation in nuclear medicine: a comparative study. Indian J Nucl Med 25:10–11
16. Marcus CS, Siegel JA (2004) NRC absorbed dose reconstruction for family member of ^{131}I therapy patient: Case study and commentary. J Nucl Med 45:13N–16N

Chapter 9
Examples of Shielding Calculations for a PET/CT Installation

Markus N. Lonsdale

9.1 Introduction

Practical radiation protection can be quite challenging. An often-encountered problem is the installation of new equipment in existing buildings that are not well suited for the specific purpose. This chapter deals with some of the radiation protection issues that came up when installing a PET/CT system.

Diagnostic radiological equipment like a CT scanner only radiates in the room where the equipment is installed. This is different for nuclear medicine. Here, the patient is the source of radiation, and patients cannot be "shielded" or constrained to a single room. Therefore, the legislation covering radiological equipment is often more rigorous than the legislation concerned with unsealed sources in nuclear medicine.

The example presented here is inspired by the circumstances at Bispebjerg Hospital in Copenhagen, Denmark. The Department of Clinical Physiology and Nuclear Medicine had received funds for acquiring a second PET/CT system. A room was allocated for installation of the scanner, and additionally, another room was to be rebuilt as uptake room for patients administered with ^{18}F-FDG. For the sake of understanding, the sample calculations here are simplified.

9.2 PET/CT System

The room for installation of the combined PET/CT system was chosen for reasons other than radiation protection. A PET/CT system is quite heavy (about 5 ton), occupies a lot of space (considerably more than 20 m^2), relies on controlled and

M.N. Lonsdale (✉)
Department of Clinical Physiology and Nuclear Medicine, Bispebjerg Hospital, 2400 Copenhagen NV, Denmark
e-mail: mlon0001@bbh.regionh.dk

S. Mattsson and C. Hoeschen (eds.), *Radiation Protection in Nuclear Medicine*, DOI 10.1007/978-3-642-31167-3_9, © Springer-Verlag Berlin Heidelberg 2013

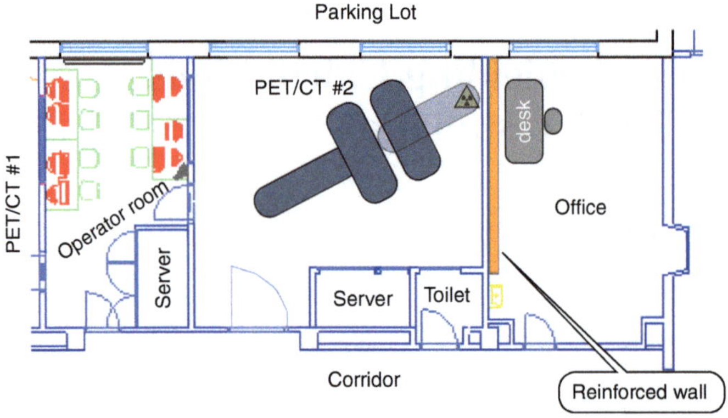

Fig. 9.1 Layout of the room for PET/CT #2 with adjacent rooms. The operator room is shared with PET/CT #1. For estimation of the dose contribution of the annihilation radiation, a radioactive source was placed in the "head" end of the patient bed (see *radioactivity symbol*) and the dose rate at the desk in the office measured. The *thick orange line* denotes the reinforced wall that was built to reduce the dose to the office staff

stable ambient environment (cooling!) while consuming a lot of electric energy (but heat dissipation is unevenly distributed over a working day), should be easily accessible for staff and patients, and much more. Radiation protection is usually secondary in negotiations with the management, architects, and entrepreneurs and often not considered beforehand. Similarly, the room to be used as uptake room was the only room of sufficient size that was available. Ideally, such a room is in close proximity to the scanner, but this was not possible here.

The second scanner was placed such that the operator room could be shared with the first scanner (see Fig. 9.1). For radiation protection purposes, two kinds of radiation (and sets of legislation) must be distinguished: 511 keV annihilation radiation from PET radionuclides (unsealed sources) and scattered radiation from the CT scanner (X-ray). In addition, relevant scenarios for occupational exposure must be selected.

The neighboring rooms of the scanner include operator room, parking lot outside the building, office, and corridor. Applicable occupancy factors for these areas can be looked up [1]. Often, neighboring rooms can be left out in shielding calculations. That can be the case if these rooms are scarcely occupied or not occupied at all. However, care must be taken that regularly occupied rooms next to the ignored rooms are taken into consideration, for example, if there is a small litter room between the radiation source and an office.

9.2.1 *Annihilation Radiation Around the Scanner Room*

As the new scanner room is neighboring an office where secretaries work throughout the day, exposure should be estimated. A simple experiment was performed, placing a vial containing 200 MBq ^{18}F-FDG in the scanner room (where the head of a patient would be placed when completely in the scanner) and measuring the dose rate at the secretary's chair—which was about 6 μSv/h (Fig. 9.1). This clearly is an overestimation of the actual dose rate, as the activity is not concentrated in the head and the patient is most of the time further away from the secretary's desk.

With an expected 1,500 patients/year, N, in that scanner, 400 MBq injected activity (having decayed to about 260 MBq, A, 1–1.5 h postinjection) and a transit time, T, of about 30 min, this would yield an estimated annual dose, D_A, of about 6 mSv—clearly unacceptable for a secretary:

$$D_A = \frac{6 \ \mu\text{Sv/h}}{200 \text{ MBq}} * T * A * N \approx 5,800 \ \ \mu\text{Sv}.$$

A reduction to 10 %, that is, accepting a maximum annual dose of 0.6 mSv, demands reinforcing the wall with 16 mm lead or 20 cm concrete [2]. The latter is actually feasible as an extra wall often can be raised without many problems (thick orange line in Fig. 9.1).

With respect to other areas around the PET/CT room, it should be noted that the distances to the patient are much larger, that is, the radiation field is much weaker or the occupancy much lower. Thus, all other areas were excluded from further calculations. The operator room also benefits from the required shielding against the scattered CT radiation, discussed in the next paragraph.

9.2.2 CT Radiation Around the Scanner Room

Legally imposed shielding requirements of a CT installation are usually fulfilled by ensuring 2 mm lead-equivalent shielding around the scanner. In this case, this was fulfilled at several places (due to sufficient wall thickness) but impractical at other places, so a more detailed investigation was required. However, measurement of scattered CT radiation with sufficient accuracy is often very difficult as the imposed radiation limits are very low.

Thus, shielding requirements were calculated based on the radiation scatter cart provided by the manufacturer and modeled transmission curves of scattered radiation from CT through lead [1]. An important parameter in this context is the CT workload, that is, the total radiation generated by the CT X-ray tube. Here, a workload of 4,000,000 mAs/year corresponding to more than 1,500 patients was assumed. The typical variety of CT protocols with the corresponding exposure settings (duration, current, and voltage) was investigated based on the experience from the first PET/CT system. For each area around the PET/CT room, occupational factors were considered.

The result was that extra shielding was only necessary between scanner and operators (2 mm lead equivalent) and at the entrance to the scanner room from the corridor (1 mm lead equivalent). During the rebuilding of the room, all doors were purchased with 2 mm lead equivalent anyway—as that option was less expensive.

9.2.2.1 Annihilation Radiation in and Around the Uptake Room

After injection with ^{18}F-FDG, patients should rest about 1 h in a relaxing environment. The uptake room, which was established for that purpose, offers space for four patients. Adjacent rooms—having functions like server room, staff toilet, etc.—are in general not occupied by persons and are thus ignored in the further calculation because of their low occupancy.

The total dose D at distance d integrating over uptake time (t_U) received from a single patient injected with activity A_0 can be estimated [2]:

$$D(t_U) = 0.092 \frac{\mu\text{Sv}\,\text{m}^2}{\text{MBq}\,\text{h}} \times A_0 \times t_U \times \frac{R_{t_U}}{d^2}.$$

Physical decay of ^{18}F is taken into account by a reduction factor R_{t_U}. As "realistic" worst-case situation we take:

- PET/CT systems with an annual workload of 2 × 1,500 patients
- Two hundred and twenty workdays, that is, about 70 patients/week
- Administered activity A_0: 400 MBq ^{18}F-FDG (administered activity for a 70 kg standard patient is normally below 300 MBq)
- One-hour rest/uptake (t_U) after injection (decay: $R_{t_U} = 0.83$)

The necessary reduction factor of a shielding arrangement limiting the radiation to a maximum dose rate $\dot{P}_{max}$ (averaged over uptake time t_U with an occupancy factor T) at distance d is

$$10.9 \times \frac{\dot{P}_{max} \times d^2}{T \times A_0 \times R_{t_U}}.$$

Likewise, the reduction factor for a given number of patients, N, over a certain time (e.g., 1 year) and a maximal dose P over the same time is given by

$$10.9 \times \frac{P \times d^2}{T \times N \times A_0 \times t_U \times R_{t_U}}.$$

In [2], attenuation coefficients of various materials reducing radiation of ^{18}F are plotted. These plots are of great help for estimation of the shielding potential of materials like lead, iron, and concrete.

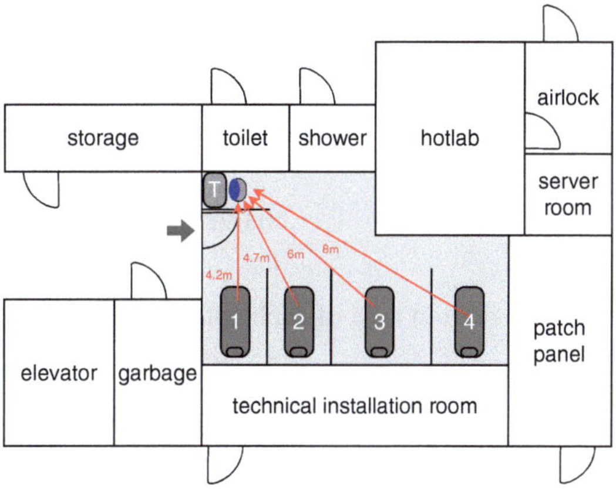

Fig. 9.2 Layout of the uptake room with adjacent rooms. The uptake room provides space for four patient beds and a small desk for a technologist doing paperwork. Distances from the middle of the patient beds to the chair of the technologist (*blue dot*) are shown. Note that adjacent rooms have low occupancy (shower, toilet, etc.) or are areas with high dose rates (hotlab)

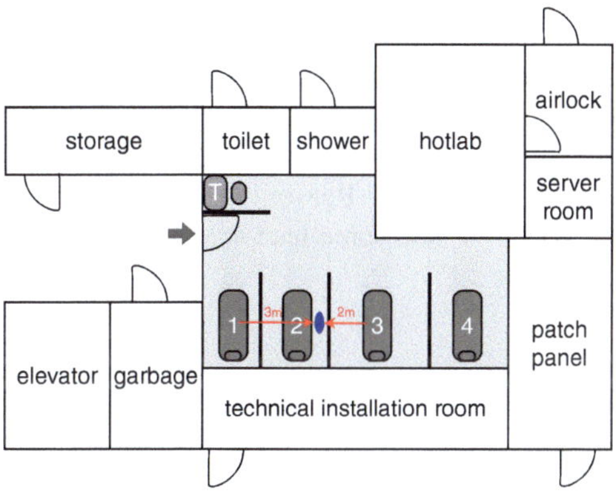

Fig. 9.3 Layout of the uptake room with adjacent rooms. Illustrated is the position of a technologist (*blue dot*) preparing patient #2 for injection, while patient #1 and patient #3 are resting in their chairs after administration of the radiopharmaceutical

With respect to the uptake room here, three relevant scenarios are considered:

(1) A technologist sitting at the table doing some registration and paperwork should be protected from the radiation of the injected patients (Fig. 9.2).
(2) A technologist preparing the injection of the next patient should be protected from radiation of the other patients (Fig. 9.3).
(3) Staff working on the floor above and below the uptake room.

9.3 Scenarios

9.3.1 Scenario 1: Technologist Working at Desk

The individual patient areas are not shielded at the feet end of the beds, so radiation from the four patients can reach the desk where the technologist can do some paperwork related to patient handling during the uptake period. While working there, staff may not be aware of working in a radiation field; thus, the area around the desk must be shielded. Constant exposure at the desk must not lead to unacceptable exposure of staff. Accepting a total dose of 0.3 mSv/month accumulated during 20 7 h working days, the average dose rate must not exceed $(0.3\,\text{mSv/month})/(20 \times 7\text{h/month}) \approx 2\ \mu\text{Sv/h}$.

Each patient contributes during the stay in the uptake room with

$$0.092 \frac{\mu\text{Sv}\,\text{m}^2}{\text{MBq}\,\text{h}} \times A_0 \times \frac{R_{t_\text{U}}}{d^2}.$$

The radiation field of each patient at distance d is thus

$$0.092 \frac{\mu\text{Sv}\,\text{m}^2}{\text{MBq}\,\text{h}} \times 400\ \text{MBq} \times \frac{0.83}{d^2} \approx 31 \frac{\mu\text{Sv}}{\text{h}} \times \frac{m^2}{d}.$$

The distance of the middle of each patient bed to the desk was measured to be 4.2, 4.7, 6, and 8 m, respectively, yielding a total dose rate of 4.5 μSv/h. A reduction to 2 μSv/h (44 %) requires 6 mm lead (or 3 cm iron) in the shielding wall at the desk.

This calculation does not take into account scattered radiation from floor, ceiling, and walls. Reasoning from simple measurements with a vial containing 4.5 GBq ^{18}F and some lead bricks, this contribution was neglected.

9.3.2 Scenario 2: Technologist Taking Care of Patient (Before ^{18}F-FDG Administration)

The distance to the closest two patients is about 2 and 3 m, corresponding to a dose rate of about 8 and 4 μSv/h, respectively, that is, 12 μSv/h in total. Reduction to 2 μSv/h (16 %) demands 13 mm lead (or 5 cm iron) to be placed in the walls between patient beds. Again, scattered radiation is ignored.

9.3.3 Scenario 3: Staff on Floor Above and Below

The floors above and below the uptake room must not be forgotten. Fortunately, below the room, there is only storage space. Above the room, however, there are several offices. Depending on national legislation, occupational exposure of non-related workers must not exceed an annual maximum dose of 0.3 mSv.

We take the same hypothetic situation as above, that is, 3,000 patients evenly distributed over four beds, all injected with 400 MBq ^{18}F-FDG. The distance from the bed to the ceiling is 2.2 m, the ceiling itself was assumed to be made of 15 cm concrete.

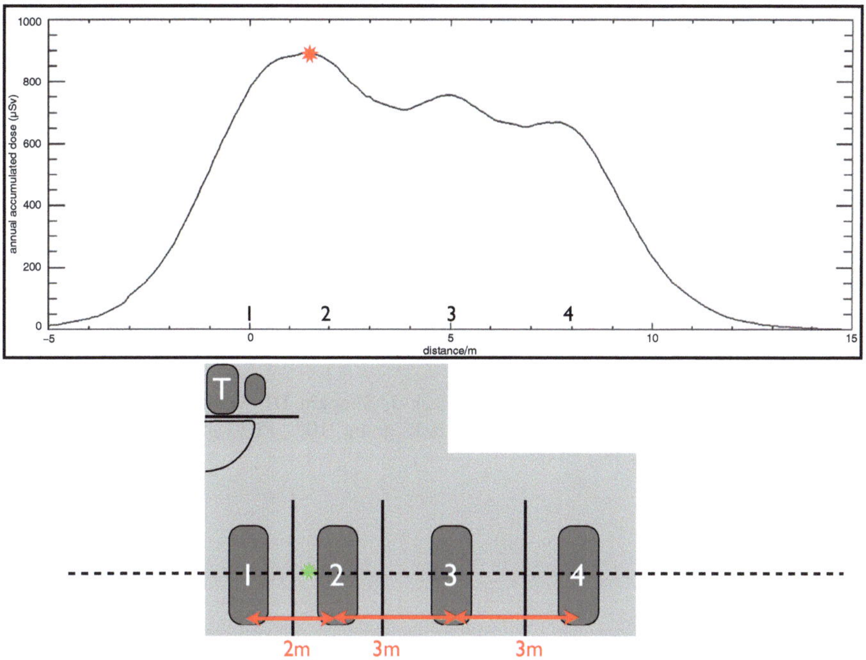

Fig. 9.4 *Upper part*: Plot of the dose profile on the floor above the uptake room. *Lower part*: Layout of the uptake room. Note that the *x*-axis of the plot corresponds to the *dashed line* in the lower part, that is, the dose rate for a person sitting on a chair in an office on the upper floor right above the *green star* will be exposed to the maximum dose (*red star*)

Considering the dose profile 2.85 m above the patient bed (in offices, people usually sit on chairs—see also [2]) and having distances between patient beds of 2, 3, and 3 m, the annual accumulated dose is just below 0.9 mSv for a person working at the "peak position" on the floor above during the same working hours as the PET technologists. This model takes into account the penetration angle of the radiation from the middle of the patient bed through concrete ceiling (Fig. 9.4).

Reducing the radiation to 1/3 (i.e., accepting an annual dose of 0.3 mSv) would take about 8 mm lead or 3–4 cm iron of shielding in the ceiling. This is not very easy to handle, and other solutions for radiation protection should be investigated. Closer inspection of the ceiling revealed that it actually is 20 cm thick, reducing transmission by an additional 50 %. Also, the location of the offices on the floor above was inspected more carefully, revealing both an additional offset in relation to the middle of the patient beds, that is, the effective distance was actually longer by about 2 m. All in all, the additional reduction in exposure was considered sufficient. It also turned out that some of the offices were used for other purposes with much lower occupancies—but that should not be relied on. It is difficult to control the use of offices belonging to other departments.

9.4 Conclusion

Practical radiation protection is often difficult to perform. This chapter has listed some examples of somewhat realistic situations in which radiation protection principles can be applied.

References

1. NCRP (2004) Structural shielding design for medical X-ray imaging facilities. NRCP report no. 147. http://www.ncrponline.org/
2. Madsen MT, Anderson JA, Halama JR, Kleck J, Simpkin DJ, Votaw JR, Wendt RE III, Williams LE, Yester M (2006) AAPM task group 108: PET and PET/CT shielding requirements. Med Phys 33(1):4–15

Part VI
Exposure of the Public

Chapter 10
Release of Patients After Radionuclide Therapy: Radionuclide Releases to the Environment from Hospitals and Patients

Sören Mattsson and Christian Bernhardsson

10.1 Introduction

Nuclear medicine involves the use of unsealed radionuclides that not only expose the investigated or treated patient but also have the potential to expose members of the public (including other patients), relatives, and caregivers. In Chap. 8, there is a description of methods for calculations and measurements of doses from patients investigated or treated by radiopharmaceuticals. The present chapter is divided into two parts. The first one deals about the release of patients from the hospital after radionuclide therapy, and in the second part, there is a discussion about radionuclide releases to the environment from hospital laboratories and through excreta from patients inside the hospital and at their homes and other places outside the hospital.

Radioiodine treatment for hyperthyroidism and thyroid cancer is the main source of exposure to the public and relatives from patients who have received unsealed radionuclides. Other radionuclides traditionally used in therapy are usually pure beta emitters (e.g., ^{32}P, ^{89}Sr, and ^{90}Y) that pose much less risk to others. Recently, a number of new therapeutic methods have come into clinical use like ^{177}Lu-octreotate (SPECT/CT) and ^{68}Ga-octreotate (PET/CT) for therapy of disseminated neuroendocrine tumors. Inoperable liver metastases are now treated with ^{90}Y-SIRS (selective internal radiation therapy) particles to create radioembolization.

After some therapeutic procedures, doses to the public, patients' relatives, and others may need to be limited, especially for children and pregnant women. After diagnostic nuclear medicine procedures, precautions for the public are rarely required.

S. Mattsson (✉) • C. Bernhardsson
Medical Radiation Physics, Department of Clinical Sciences Malmö, Lund University, Skåne University Hospital, 205 02 Malmö, Sweden
e-mail: soren.mattsson@med.lu.se

S. Mattsson and C. Hoeschen (eds.), *Radiation Protection in Nuclear Medicine*,
DOI 10.1007/978-3-642-31167-3_10,

Doses to other people from patients who have received radioiodine therapy are predominantly due to external exposure. For adult family members, contamination is much less important than external exposure. On the other hand, it is important to avoid contamination of children and pregnant women due to the sensitivity of the fetal thyroid and the thyroid glands of children.

A common question after a radionuclide therapy is whether the patient has to be hospitalized or not and which release criteria should be applied.

The releases of radionuclides to the environment are dominated by activity in patient excreta (urine and feces) and consist to a minor extent of releases from hospital laboratories. In some countries, there is an ongoing discussion between hospitals and regulatory bodies whether storage tanks are needed at the hospital or not to delay, and thus decrease, the release of the activity.

10.2 Release of Patients After Radionuclide Therapy

There is still no harmonization of national regulations and not either of ICRP, IAEA, or EU guidelines. Current release criteria for patients treated with ^{131}I-, ^{32}P-, ^{89}Sr-, ^{153}Sm-, and ^{90}Y-compounds are given in Table 10.1 [1].

The above-mentioned release criteria are more and more questioned. The basic aim of such criteria is that members of the public should remain within dose limits and that comforter and caregivers (often family members) should be subject to dose constraints. Comforters and caregivers are defined as "those who *knowingly* and *willingly* help, (other than as part of their occupation) with the care and support of the patient."

The dose limits for the public (which includes family, other than comforters and caregivers), friends and acquaintances, visitors, casual visitors, work colleagues, and those encountered socially or while traveling is 1 mSv/year. ICRP [3], IAEA [2, 4], EU [5], and national authorities all agree on that.

ICRP Publication 73 [3] recommends that dose constraints should be established, but the ICRP has not yet recommended values for such constraints, but a value in the region of a few mSv/episode is likely to be reasonable. The constraints are not to be used rigidly. For example, higher doses may well be appropriate for the parents of very sick children.

For comforters, BSS [2] recommends a dose constraint of 5 mSv during the period of patient's treatment. IAEA [1] has recently proposed numerical values for dose constraints for various personal groups (see Table 10.2).

There are several reasons to limit the hospitalization of patients: Patients do not want to stay longer than necessary. Patients need privacy. There are also financial reasons for the health care systems and for the society to keep the hospitalization as short as possible. However, many countries tend to enforce practices *forgetting the basis* and stick to quantities of retained radioactivity in the patient's body or dose rate at the defined distance from the patient [1].

Table 10.1 Release criteria for patients after radionuclide therapy (after [1])

	Max retained activity (MBq) in the body at release				
Country	^{131}I	^{32}P	^{89}Sr	^{153}Sm	^{90}Y
Finland	800				
Germany	75 (250)				
EU (most countries)	400–600				
Japan	500		200		1,200
USA	1,200			26,000	
Sweden	600	1,200			1,200
Australia	600	1,200	300	4,000	4,000
Eur thyr ass	800				
BSS	1,100				

The revised (interim) BSS [2] does not give numerical values

Table 10.2 Proposed dose constraints/episode for different categories of persons/caregivers [1, 6]

Type of person/caregiver	Dose constraint	Reason
Third person (not carer)	0.3 mSv/episode	A fraction of the dose limit for the public
Family and close friends		
Pregnant women	1 mSv/year	Protection of the unborn child
Children up to 2 years old	1 mSv/year	Close physical contacts with parents
Children between 3 and 10 years	1 mSv/episode	Same risk as for 0–2 years old and unborn
Children older than 10 years and adults up to 60 years old	3 mSv/episode	2–3 times lower risk than for younger children
Adults older than 60 years	15 mSv/episode	3–10 times lower risk than for 10–60 years old

ICRP [6] does not explicitly require hospitalization, but IAEA [7] writes that it is not recommended to let the patient return home immediately. Instead, he or she should be kept in the hospital for a period between some hours and some days. The EU recommendations however state that treatment of thyroid cancer using radioiodine should only be performed on inpatients [5].

In USA, the Nuclear Regulatory Commission (NRC) amended its regulations from an activity-based limit to a dose-based limit in 1997. The new regulation is based on the maximally exposed individual not being likely to exceed an effective dose equivalent of 5 mSv [8].

Concerning exposure of persons in home environment, actual measurements of families or caregivers who are able to follow radiation protection precautions show that doses rarely approach or exceed the few mSv/episode, which is the ICRP recommended dose constraint [9].

ICRP Publication 94 [6] states that the decision to hospitalize or release a patient should be determined on individual basis. It should consider factors such as residual activity in the patient, patient's wishes, occupational and public exposure, family considerations, cost, and environmental aspects.

Table 10.3 Proposed guidelines for radioiodine patients to restrict dose to 1 mSv to coworkers and family [1, 6]

Activity MBq	Patient type	Time off work (*d*)		Time to sleep apart and restrict contact with partner (*d*)		Time to restrict contact with child <2 years of age (*d*)		Time to restrict contact with child 2–5 years of age (*d*)		Time to restrict contact with child 5–11 years of age (*d*)	
200	Hyperthyroidism	0		15		15		11		5	
400	Hyperthyroidism	3		20		21		16		11	
600	Hyperthyroidism	6		24		24		20		14	
800	Hyperthyroidism	8		26		27		22		16	
		F-up	Abl	F-up	Abl	F-up	Abl	F-up	Abl	F-up	Abl
1,850	Cancer	1	3	3	16	4	16	3	13	2	10
3,700	Cancer	2	7	4	20	4	20	4	17	3	13
5,550	Cancer	2	10	4	22	5	22	4	19	3	16
7,400	Cancer	2	12	5	23	5	24	4	21	4	17

F-up means cancer follow up patients; *Abl* means ablation patients

Table 10.4 Residual activities, dose rates, and time periods for which instructions must be followed according to EU radioiodine recommendations [1]

Residual activity (MBq)	Corresponding effective dose rate at 1 m from patient (μSv/h)	Period for which instructions must be followed
<60	<3	24 h
<100	<5	4 days
<200	<10	1 week
<400	<20	2 weeks
<800	<40	3 weeks

A single model for release criteria would not be appropriate optimization. It is recommended that release of patients should be based on an individual basis (rather than retained activity and the worst-case scenario). It is also recommended that where there are many nearby countries, a uniform or similar approach to releasing patients should be developed. The IAEA Safety Report Series No. 63 [1] is an attempt to help resolve diversity of international practice.

Both ICRP [6] and IAEA [1] recommend:

- A dose constraint of a few mSv/episode for caregivers (Table 10.2)
- A dose limit of 1 mSv/year (as for the public) for infants, young children, and casual visitors

Tables 10.3, 10.4, and 10.5 may be helpful to reach these goals.

Table 10.5 Periods of restrictions after administration of ^{131}I [1, 6]

Activity (MBq)	All close contact with children or pregnant women (*d*)	Extended contact with children or pregnant women (*d*)	Do not share bed (*d*)	Avoid prolonged close contact with others (*d*)
30–400	9	21	0	0
400–600	12	25	4	0
600–800	14	27	8	1

10.2.1 Patient Travel

Patients traveling after radioiodine therapy rarely present hazard to other passengers if travel time is a few hours and restrictions following release of patients should focus on infants and children. ICRP [6] and IAEA [1] recommend that the public dose limit of 1 mSv/year shall apply to infants, children, and casual visitors.

Environmental or other radiation-detection devices are able to detect patients who have had radioiodine therapy for several weeks after treatment. Personnel at borders between countries operating such detectors should be specifically trained to identify and deal with nuclear medicine patients. Records of the therapies with unsealed radionuclides should be maintained at the hospital and given to the patient along with written precautionary instructions.

10.2.2 Immediate Medical Intervention, Postmortem Examinations, and Burial

If the patient has had radiotherapy with unsealed radionuclides in the last few months, special precautions may be required. Please see Chap. 8.

10.3 Releases to the Environment

The radioactive waste from the nuclear medicine activities is in the form of liquid stock solutions, blood and urine samples, contaminated syringes, injection needles, gloves, etc. This waste is normally kept at the hospital for a sufficient period of time for the radionuclides to decay below the limits authorized by the national authority—the so-called "delay and decay" technique. Thereafter, most of it can be handled and discharged as nonradioactive waste. The residue of long-lived radionuclides and/or radionuclides of high activity has to be taken care of by specialized companies for longer time storage.

For the controlled discharges to the sewer system, there are detailed rules, and the goal is to limit the dose contribution to a person who represents the members of the public who can be expected to get the highest doses. In Sweden, this value is

Table 10.6 Estimated releases to the environment in Sweden from various sources during 2003 [11]

Radionuclide	Nuclear industries (GBq)	Medical care (GBq)	Research and industry (GBq)
^{3}H	82,000		1,300
^{14}C	4,600		20
^{18}F		560	
^{32}P		58	53
^{35}S			26
^{51}Cr	7.6		5
^{54}Mn	2.9		
^{60}Co	32		
^{65}Zn	0.8		0.01
^{89}Sr		20	
^{90}Sr	3.6		
^{99m}Tc	0.3	34,000	3
^{110m}Ag	1.8		
^{125}I			5
^{131}I	0.8	2,000	0.1
^{137}Cs	11		
^{153}Sm		730	

For the nuclear industries, data refer to the SSI reported measured emissions in 2003, and medical care refers to the SSI reported *administered* amount in 2003. Emissions from research and industry are estimated on the basis of ongoing contacts with the licensees

10 μSv/year [10]. This means that the allowed releases for ^{131}I are only 10 MBq/month (max 1 MBq each time) and for ^{99m}Tc 100 MBq/month (max 10 MBq each time). All releases should be done from one specific sink at the hospital. If higher releases are needed, realistic calculations have to be carried out and documented to show that the limit 10 μSv/year can be kept.

The radionuclide releases from hospitals are however dominated by activity in the excreta from patients, who have been given radiopharmaceuticals for therapy or diagnostic purposes. Excretion of radionuclides from these patients, when back at home, will also add to the total activity released into the environment.

The estimated radioactive releases from nuclear industries, hospitals, research laboratories, and industries in Sweden during 2003 are shown in Table 10.6 (from Andersson et al. [11]).

For the column medical care, information about the *administered* activity is given. Only a fraction of the activity given to patients will reach the environment. Approximate proportion of administered activity (until total decay) discharged to drains can be estimated using information in Table 10.7 [1].

Technetium-99 m dominates the discharges to the environment from excreta of nuclear medicine patients, but its short half-life limits its importance. The second largest discharges, ^{131}I, can be detected not only in wastewater systems and sewage treatment plants but also in the environment. The transfer of the water phase in the treatment plant is comparatively short (around 5–7 h). For the solid phase, it takes longer time, normally 2–3 weeks. Figure 10.1 [12] demonstrates the results from an

Table 10.7 Estimated relative discharges of the activity of various radiopharmaceuticals [1]

Radionuclide and form		For treatment of	Relative discharge
^{198}Au	Colloid	Malignant disease	0
^{131}I	Iodine	Hyperthyroidism	50
^{131}I	Iodine	Thyroid cancer	80–90
^{131}I	MIBG	Endocrine tumors	90
^{32}P	Phosphate	Polycythemia vera	40
^{89}Sr	Chloride	Bone metastases	90
^{90}Y	Colloid	Arthritis joints	0
^{90}Y	Antibody	Tumors	10
^{169}Er	Colloid	Arthritis joints	0

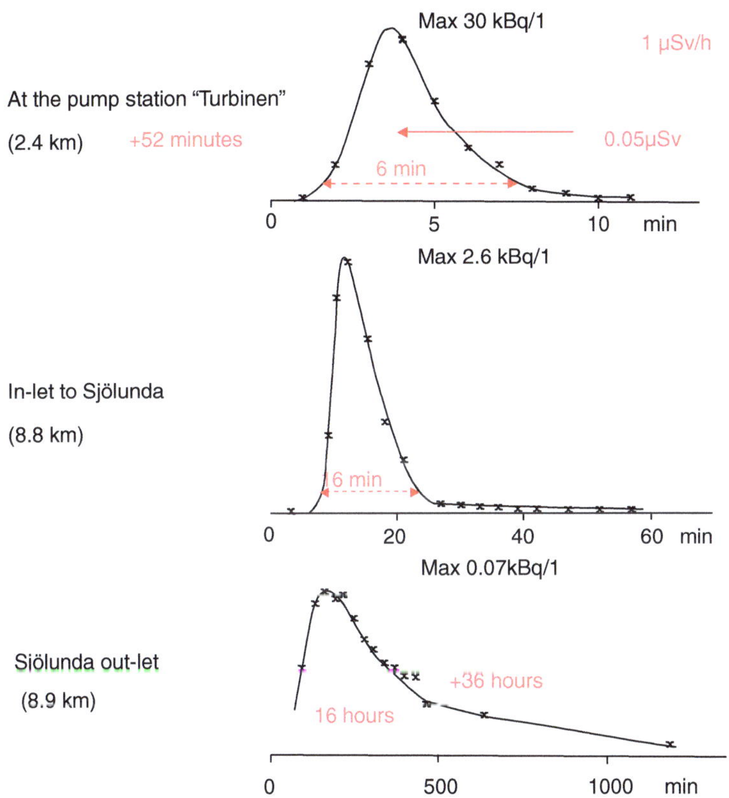

Fig. 10.1 Dose rates measured at different times and distances from Malmö General Hospital after a controlled release of 3 GBq of ^{131}I in 72 l of urine followed by 500 l of water [12]

experiment using an instantaneous, controlled input of urine earlier collected from a number of patients treated with ^{131}I into a sink in the hospital. Measurements of dose rate as a function of time were done along the wastewater pipe all the way to the treatment plant (Sjölunda). The activity concentration in the digested sewage

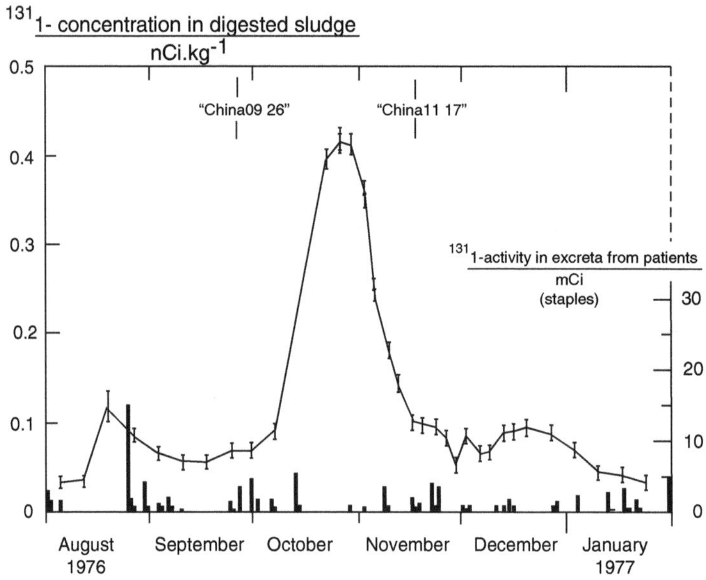

Fig. 10.2 ^{131}I-activity concentration in digested sludge (left *y*-axis) and estimated daily release of ^{131}I from patients (right *y*-axis). The ^{131}I-contribution from the two indicated Chinese nuclear weapons tests arrived 2 weeks later to Malmö [13]

sludge was also studied, and the influence of individual ^{131}I-therapies was easily seen with a delay of 2–3 weeks (Fig. 10.2). During the period of measurements, there was incidentally also a deposition through rain of ^{131}I from two Chinese nuclear tests, which dominated the ^{131}I content in the sludge. Based on information on released activity from the hospital and the dynamics of the sewage system, the conclusion was that radionuclides released into modern sewage systems are likely to result in doses to sewer workers and the public that are well below the mentioned dose limits of 10 μSv/year. First with much higher frequency of therapies (say 10/week), more careful estimates have to be done.

Therefore, storing patients' excreta after therapy appears to have minimal benefit. Holding tanks are not practical because of costs and exposure of hospital staff and because significant proportions of discharges occur after patients have left the hospital (more and more patient are now treated as outpatients). ICRP recommendations [6] do not require urine to be stored.

The releases to water bodies (rivers or coastal waters) may result in measurable activity concentrations in, e.g., algae, but the levels are very low and constitute no radiation protection problem.

Use of sewage sludge for energy forests, golf courses, and as landfill presents no health concerns because sufficient decay occurs before it can reach the public. Direct incineration of sludge could however be an unnecessary source for human exposure and should be avoided until measurements on the sludge have been done.

10.4 Summary

The decision to hospitalize or release a patient after therapy with radiopharmaceuticals should be determined on an individual basis. It should consider factors such as residual activity in the patient, patient's wishes, occupational and public exposure, family considerations, cost, and environmental aspects and meet the requirements of dose constraint of a few mSv/episode for caregivers and be planned so that infants, young children, and casual visitors are subjected to the public dose limit of 1 mSv/year.

Concerning releases through patient excreta, estimates of effective doses to sewer workers, wastewater treatment operators, and sludge press workers have given very low values. Therefore, storing patients' excreta after therapy appears to have minimal benefit. Holding tanks are not practical because of costs and exposure of hospital staff and because significant proportions of discharges occur after patient release (more and more patient treated as outpatients).

References

1. International Atomic Energy Agency, with contributions from International Commission on Radiological Protection (2009) Release of patients after radionuclide therapy. IAEA Safety Report Series No. 63, Vienna. http://www-pub.iaea.org/MTCD/publications/PDF/pub1417_web.pdf
2. International Atomic Energy Agency (2011) Radiation protection and safety of radiation sources: international basic safety standards. Interim edition. Safety Standards Series No. GSR Part 3 (interim). IAEA, Vienna
3. International Commission on Radiological Protection (1996) Radiological protection and safety in medicine. ICRP Publication 73. Ann ICRP 26(2)
4. International Atomic Energy Agency (1996) International basic safety standards for protection against ionizing radiation and for the safety of radiation sources. IAEA Safety Series No. 115. IAEA, Vienna
5. European Commission (1998) Radiation protection following ^{131}I-therapy. Exposures due to out-patients or discharged in-patients. Radiation Protection 97. Directorate General for Environment Nuclear Safety and Civil Protection, Brussels
6. International Commission on Radiological Protection (2004) Release of patients after therapy with unsealed radionuclides. ICRP Publication 94. Ann ICRP 34(2)
7. International Atomic Energy Agency, Pan American Health Organization, World Health Organisation (2002) Radiological protection for medical exposure to ionizing radiation. IAEA safety standards series no. RS-G-1.5. IAEA, Vienna
8. National Council on Radiation Protection and Measurements (2011) Management of radionuclide therapy patients. NCRP report no. 155, NCRP
9. Barrington SF, O'Doherty MJ, Kettle AG, Thomson WH, Mountford PJ, Burrell DN, Farrell RJ, Batchelor S, Seed P, Harding LK (1999) Radiation exposures of families of outpatients treated with radioactive iodine (iodine-131) for hyperthyroidism. Eur J Nucl Med 26:686–692
10. Swedish Radiation Safety Authority (2010) Radiation safety authority's regulations on radioactive waste and discharges from activities with unsealed sources. SSMFS 2010:2. SSM, Stockholm (in Swedish)

11. Andersson P, Carlsson M, Falk R, Hubbard L, Leitz W, Mjönes L, Möre H, Nyblom L, Söderman A-L, Yuen Lasson K, Åkerblom G, Öhlén E (2007) The radiation environment in Sweden. Report SSI 2007:02 2007. ISSN 0282-4434 (in Swedish)
12. Mattsson S, Erlandsson B (1981) Medicinskt använda radionuklider i avloppsvatten och rötslam. SSI-Report P105 (in Swedish)
13. Erlandsson B, Mattsson S (1978) Medically used radionuclides in sewage sludge. Water Air Soil Pollut 9:199–206

Part VII
Rules of Thumb for Radiation Protection in Nuclear Medicine

Chapter 11
Rules of the Thumb and Practical Hints for Radiation Protection in Nuclear Medicine

Sören Mattsson and Martin Andersson

11.1 Introduction

In radiation protection, as in many other fields, there is a need for rapid estimates of doses and consequences.

The US Public Health Service already 60 years ago developed its first Radiological Health Handbook. The 1970 edition of this handbook, which was compiled and edited by the Bureau of Radiological Health and the Training Institute Environmental Control Administration, has been a classical handbook containing a number of very valuable rules of the thumb and practical advices related to occupational radiation protection. Later, the UK Society of Radiological Protection has put together valuable information on its homepage under the name of rules of thumb and practical hints (http://www.srp-uk.org/resources/rules-of-thumb-a-practical-hints/103-rules-of-thumb-and-practical-hints). Similar collections have since 1992 been available from The Health Physics and Radiological Health Handbook (http://www.epa.gov/radiation/docs/wipp/08-0442-attach-3.pdf). Several textbooks also contain valuable information of that kind. In relation to the protection of patients, the International Commission of Radiological Protection (ICRP) has published a series of documents relating the administered activity to organ doses and effective doses.

The aim of this chapter is to extract existing information of interest in nuclear medicine and to add new such information.

S. Mattsson (✉) • M. Andersson
Medical Radiation Physics, Department of Clinical Sciences Malmö, Lund University, Skåne University Hospital, 205 02 Malmö, Sweden
e-mail: soren.mattsson@med.lu.se

S. Mattsson and C. Hoeschen (eds.), *Radiation Protection in Nuclear Medicine*,
DOI 10.1007/978-3-642-31167-3_11, © Springer-Verlag Berlin Heidelberg 2013

11.2 Occupational Exposure in Nuclear Medicine

11.2.1 Basic Principles of Radiation Protection

When working with radionuclides, the fundamental principles are:
1. Minimize the time spent in vicinity to the radionuclide—time.
2. Try to optimize the distance to the radionuclide—distance.
3. Work behind proper shielding—shielding.

It is important to practice the working procedures to optimize the radiation protection, e.g., with nonactive material, in advance.

11.2.1.1 Simple Approximative Relations

1. The range, R, of beta particles in g/cm^2 is approximately equal to the maximum energy, E, in MeV divided by 2 (i.e., $R = E/2$); this means that a 1 MeV electron has an approximately 0.5 cm range in water [1, 2].
2. The range of beta particles in air is about 3½ m/MeV [1].
3. It requires a beta particle of at least 70 keV to penetrate the protective layer of skin, 0.07 mm thick [2].
4. It requires an alpha particle of at least 7.5 MeV to penetrate the protective layer of skin, 0.07 mm thick [2].
5. The activity of any radionuclide is reduced to <1 % after seven half-lives [2].
6. For gamma energies between 60 keV and 1.5 MeV, the dose rate from a source A MBq and total energy emission per disintegration of E MeV $= 0.14\,AE$ μGy/h at 1 m [1].
7. For material with a half-life >6 days, the change in activity in 24 h will be <10 % [2].
8. For proton energies, E, between few MeV to 200 MeV, the approximate range, R, in meters in air is $R = (E/9.3)^{1.8}$ [2].

11.2.2 Shielding

1. For beta shielding, low atomic number materials give you less bremsstrahlung production but also less bremsstrahlung attenuation. The optimum shielding solution requires a balance.
2. The lower the Z of the material, the worse it scatters X- and gamma radiation [3].
3. At shielding of high gamma energies, it is mass/unit area that is important, not atomic number.

Table 11.1 Half-value layers for different radionuclides in lead [4]

Radionuclide	HVL Pb (mm)
^{18}F	4.0
^{67}Ga	1.4
^{99m}Tc	0.3
^{111}In	0.7
^{123}I	0.4
^{125}I	0.02
^{131}I	3.0

4. For high-energy gamma radiation, the denser the shielding material, the lower the mass of the shield for a source container. Density is more important than Z.
5. An approximate HVL for 1 MeV neutrons is ~3 cm of polythene or water, ~7 cm for 5 MeV neutrons [1].
6. Half-value layers for different radionuclides in lead [4] are given in Table 11.1.

11.2.3 Contamination

11.2.3.1 Some Basic Guidelines to Minimize the Exposure from Contamination

1. Laboratory gloves minimize the skin dose from electrons (a 0.15–0.25 mm protective glove lowers the dose rate 60 times to the basal-cells from a ^{99m}Tc-nuclide) [5].
2. Change gloves frequently (there is always some amount of contamination) and avoid touching "clean areas" or touching your face.
3. If an accidental wound occurs, stop procedure; wash and patch wound before continuing.
4. After five hand washes with water and soap, only 2 % of the initial activity is remaining on the skin surface [4].
5. Use protective goggles if there is a risk of splash.

11.2.3.2 When a Contamination Occurs There Are Some Basic Advices to Follow

1. Change clothes and shoes and wash unprotected skin and clean surfaces close to the contaminated area (minimizes the spreading of contamination).
2. Restrict the contaminated areas (limit the access for unauthorized).
3. Get adequate information about the contaminated area before starting the decontamination.

11.2.3.3 Rough Absorbed Dose Estimates

For normal health physics purposes, the skin dose per activity per unit area can be approximated as 1 mSv/h per Bq/cm^2 [1].

11.2.3.4 Shielding of Syringes and Radionuclide Solutions

If an injection of radionuclides is made manually, use of a syringe shield will reduce the dose to hand and fingers.

1. A syringe shield reduces the dose with 50–85 % to hands and fingers for ^{99m}Tc and 25 % for PET substances [6].

11.2.4 External Exposure

External Exposure from 1 MBq of Various Radionuclides at Contact and at a Distance of 1 m [4] (Tables 11.2 and 11.3)

11.2.4.1 Lead Apron or Not in Nuclear Medicine?

In nuclear medicine, where the energies of the ambient radiation are much higher than in radiography, the lead apron is of limited use and is often considered too restrictive for day long wear.

1. A 0.25 mm lead apron will provide a dose reduction of about only about 59 % for ^{99m}Tc (140 keV) [7].
2. A 0.5 mm lead apron weighs about 8.5 kg and will provide a dose reduction of about 76 % [8].
3. For higher energies, such as ^{131}I (360 keV) and ^{18}F (511 keV), an apron is of little use [8].

11.2.4.2 Occupational Exposure in Computed Tomography

Computed tomography (CT) is more and more used in connection with SPECT and PET. The CT is primarily used to improve the attenuation correction of the SPECT and PET images and secondarily as a diagnostic modality.

1. Scatter radiation from a patient depends on patient size and tube voltage, but optimizing the distance to the patient sufficiently lowers the dose rate.
2. Standing behind the CT gantry gives the optimal shielding position in vicinity to the patient [9].

11.2.4.3 Lead Apron or Not in CT?

From a CT, the mean photon energy is approximately 1/3 of the nominal tube voltage. A CT with a tube voltage between 90 and 120 keV generates mean photon energies of 30–40 keV.

Table 11.2 External exposure from 1 MBq at contact [4]

Radionuclide	μSv/h at contact with 5 ml syringe	μSv/h at contact with 10 ml glass vial	Use
^{3}H	<1	0	In vitro
^{14}C	<1	0	In vitro
^{32}P	23,900	0.0054	In vivo/therapy
^{35}S	<1	0	In vitro
^{125}I	620	0.014	In vitro
^{18}F	2,880	0.16	Diagnosis
^{67}Ga	402	0.025	Diagnosis
^{111}In	1,220	0.072	Diagnosis
^{99m}Tc	354	0.022	Diagnosis
^{123}I	605	0.034	Diagnosis
^{89}Sr	16,400	0.00018	Therapy
^{90}Y	43,500	0.071	Therapy
^{131}I	1,130	0.063	Therapy
^{153}Sm	241	0.015	Therapy

Table 11.3 Mean dose rate at 1 m from a patient after administration of a radiopharmaceutical [4]

Investigation/treatment	Radiopharmaceutical	Administered activity, MBq	μSv/h
Bone scan	^{99m}Tc-MDP	740	5
Cardiac	^{99m}Tc-MIBI	740	5
Tumor imaging	^{18}F-FDG	370	55
Neuroreceptor imaging	^{123}I-ioflupane (Datscan)	111	2
Neuroendocrine tumor imaging	^{111}In-octreotide	111	2
Thyroid cancer therapy	^{131}I (iodide)	7,400	200
Lymphoma therapy	^{90}Y-Zevalin	900	1

1. A 0.25 mm lead apron will provide a dose reduction of about 94.6 % at 70 kV and 85 % at 100 kV [10].
2. A 0.50 mm lead apron will provide a dose reduction of about 99.1 % for 70 kV and 95 % at 100 kV [10].

If it is necessary to assist in the CT room during a procedure, an apron shall be used. If there is not enough aprons (even if it is never recommended) shielding behind a human can act as a temporary protection.

11.3 Patient Exposure

See Table 11.4a, b.

Table 11.4 (**a** and **b**) Effective dose/unit activity administered E/A_0 (for adults)

Radiopharmaceutical	Effective dose per unit activity administered (E/A_0), mSv/MBq	Ref.	"Typical activity" used (A_0), MBq	Effective dose per investigation, mSv
(a)				
^{3}H-neutral fat, free fatty acids	0.22	80	0.1	0.022
^{11}C-acetate	0.0035	106	500–1,000	1.8–3.5
^{11}C-amino acids (generic model)	0.0056	106	200–400	1.1–2.2
^{11}C-choline	0.0047	TG	400	1.9
^{11}C-brain receptor subst. (generic model)	0.0043	106	400	1.7
^{11}C-methionine	0.0084	106	200–400	1.7–3.4
^{11}C-thymidine [methyl-^{11}C] thymidine	0.0035	80	100–1,000	0.35–3.5
^{11}C-thymidine [2-^{11}C] thymidine	0.0027	80	400	1.1
^{11}C (realistic maximum model)	0.011	106	400	4.4
^{14}C-neutral fat, free fatty acids	2.1	80	0.1	0.21
^{14}C-urea (normal/*Helicobacter* pos.)	0.031/0.081	80	0.1–0.2	0.003–0.016
^{15}O-water	0.0011	106	400	0.44
^{18}F-amino acids (generic model)	0.023	106	400	9.2
^{18}F-brain receptor subst. (generic model)	0.028	106	400	11
^{18}F-choline	0.019	TG	400	7.6
^{18}F-FDG	0.019	106	250–400	4.8–7.6
^{18}F-FET	0.017	TG	400	6.8
^{18}F-fluoride	0.024	80	400	9.6
^{18}F-L-dopa	0.025	106	400	10
^{18}F-FLT	0.016	TG	400	6.4
^{51}Cr-EDTA	0.0020	80	3–4	0.006–0.008
^{67}Ga-citrate	0.10	80	150	15
^{75}Se-HCAT	0.69	80	0.02; 0.4[a]	0.01; 2.8[a]
^{82}Rb-chloride	0.0017[b]	TG	3,000	5.1
^{99m}Tc-apticide	0.0047	106	750	3.5
^{99m}Tc-colloids (large)	0.0094	80	80–500	0.75–4.7
^{99m}Tc-colloids (small), intratum. inj.	0.0012	106	50–100	0.060–0.12
^{99m}Tc-DMSA	0.0088	80	100	0.88
^{99m}Tc-DTPA	0.0049	80	300	1.5
^{99m}Tc-EC	0.0063	106	500	3.2
^{99m}Tc-ECD	0.0077	106	500	3.9
^{99m}Tc-furifosmin (rest/exercise)	0.010/0.0089	106	300–900	2.7–9.0

(continued)

Table 11.4 (continued)

Radiopharmaceutical	Effective dose per unit activity administered (E/A_0), mSv/MBq	Ref.	"Typical activity" used (A_0), MBq	Effective dose per investigation, mSv
^{99m}Tc-HM-PAO	0.0093	80	500–1,000	4.7–9.3
^{99m}Tc-IDA derivatives	0.017	80	150	2.6
^{99m}Tc-MAA	0.011	80	100–200	1.1–2.2
^{99m}Tc-MAG3	0.0070	80	70–200	0.49–1.4
^{99m}Tc-markers, nonabsorbable (fluids/solids) per os	0.019/0.024	80	10–40	0.19–0.96
(b)				
^{99m}Tc-MIBI (rest/exercise)	0.0090/0.0079	80	300–900	2.4–8.1
^{99m}Tc-monoclonal antibodies: intact ab/F(ab′)$_2$-fragm/F (ab′)-fragm	0.0098/0.0097/0.011	106	750	7.3–8.3
^{99m}Tc-pertechnetate, without blocking	0.013	80	100–800	1.3–10
^{99m}Tc-pertechnetate, with blocking	0.0042	80	100–800	0.42–3.4
^{99m}Tc-phosphates and phosphonates	0.0057	80	600	3.4
^{99m}Tc-RBC	0.0070	80	800	5.6
^{99m}Tc-pertechnegas/-Technegas	0.012–0.015	80	30	0.36–0.45
^{99m}Tc-tetrofosmin (rest/exercise)	0.0080/0.0069	106	600	4.1–4.8
^{99m}Tc-WBC	0.011	80	200	2.2
^{111}In-monoclonal antibodies: intact ab/F(ab′)$_2$-fragm/F (ab′)-fragm	0.22/0.20/0.20	106	200	40–44
^{111}In-octreotide	0.054	106	170	9.2
^{123}I-iodide, 35 % thyroid uptake	0.22	80	2–20	0.44–4.4
^{123}I-iodide, after ablation, 1 % thyroid uptake	0.02	53	200–400	4–8
^{123}I-fatty acids (BMIPP/IPPA)	0.016	106	150	2.4
^{123}I-brain receptor subst. (generic model)	0.050	106	180	9.0
^{123}I-MIBG	0.013		400	5.2
^{123}I-monoclonal antibodies: intact ab/F(ab′)$_2$-fragm/F (ab′)-fragm	0.026/0.019/0.017	106	200	3.4–5.2
^{131}I-iodide, 35 % thyroid uptake	24	80	0.2	4.8
^{131}I-iodide, after ablation, 1 % thyroid uptake	0.3	53	4	1.2

(continued)

Table 11.4 (continued)

Radiopharmaceutical	Effective dose per unit activity administered (E/A_0), mSv/MBq	Ref.	"Typical activity" used (A_0), MBq	Effective dose per investigation, mSv
^{131}I-iodohippurate	0.052	80	0.2	0.010
^{131}I-monoclonal antibodies: intact ab/F(ab′)$_2$-fragm/F (ab′)-fragm	0.42/0.14/0.11	106	200	22–84
^{201}Tl-chloride	0.14	106	80	11

As references for the effective dose per unit activity administered, the number of the relevant ICRP publication [11–14] is given. TG means preliminary estimate by the ICRP task group. In order to give the reader a view of the magnitude of the effective dose, information on "typical activity" is given: these are activities that are currently used in some places, and they should not be interpreted as a recommendation of the optimal activity to be administered. The ranges of "typical" activities used are in some cases explained by different indications for the investigations [15]

[a]Measured with whole-body counter and gamma camera, respectively

[b]Interim values [16, 17]

References

1. The Society for Radiological protection. Rules of thumb and practical hints, http://www.srp-uk.org/resources/rules-of-thumb-a-practical-hints/103-rules-of-thumb-and-practical-hints
2. US Department of Health, Education and Welfare (1970) Radiological health handbook. Revised edition, p 458
3. Shleien B (ed) (1992) The Health physics and radiological health handbook. http://www.epa.gov/radiation/docs/wipp/08-0442-attach-3.pdf
4. EANM (2007) Best practice in nuclear medicine. Part 2. A technologist's guide. http://www.schreisupport.com/Schreisupport/EANM%20Guidelines/tech_best_practice2.pdf
5. Bjurman B, Ahlgren L, Mattsson S (1982) Radiation dose to staff handling $^{99}Tc^{m}$ in hospitals. In: Proceedings of the society for radiological protection: third international symposium. Inverness, Scotland
6. Vanhavere F, Carinou E, Donadille L, Ginjaume M, Jankowski J, Rimpler A, Sans Merce M (2008) An overview on extremity dosimetry in medical applications. Radiat Prot Dosim 129:350–355
7. Warren-Forward H, Cardew P, Clack L, McWhirter K, Johnson S, Wessel K (2007) A comparison of dose savings of lead and lightweight aprons for shielding of 99 m-technetium radiation. Radiat Prot Dosim 124:89–96
8. Ahmed S, Zimmer A, McDonald N, Spies S (2007) The effectiveness of lead aprons in reducing radiation exposures from specific radionuclides. J Nucl Med 48(2):470
9. Harvey SB (2010) Measurement of scatter radiation from the GE Infinia Hawkeye 4 SPECT/CT system. Aust Phys Eng Sci Med 33:193–197
10. Christodoulou EG, Goodsitt MM, Larson SC, Darner KL, Satti J, Chan HP (2003) Evaluation of the transmitted exposure through lead equivalent aprons used in a radiology department, including the contribution from backscatter. Med Phys 30:1033–1038
11. ICRP (1987) Radiation dose to patients from radiopharmaceuticals. ICRP Publ. 53. Ann ICRP 18(1–4)
12. ICRP (1998) Radiation dose to patients from radiopharmaceuticals. Addendum 2 to ICRP Publication 53. ICRP Publ 80. Ann ICRP 28(3)
13. ICRP (2007) The 2007 recommendations of the international commission on radiological protection. ICRP Publication 103. Ann ICRP 37(2–4)

14. ICRP (2008) Radiation dose to patients from radiopharmaceuticals A third amendment to ICRP Publication 53. ICRP Publ 106. Ann ICRP 38(1–2)
15. Mattsson S, Johansson L, Leide-Svegborn S, Liniecki J, Nosske D, Riklund K, Stabin M, Taylor D (2011) Current activities in the ICRP concerning estimation of radiation doses to patients from radiopharmaceuticals for diagnostic use. J Phys Conf Ser 317(1–6):012008. doi:10.1088/1742-6596/317/1/012008
16. Stabin MG (2010) Proposed revision to the radiation dosimetry of ^{82}Rb. Health Phys 99 (6):811–813
17. Leggett RW, Williams LR, Eckerman KF (1996) A blood circulation model for reference man. In: Proceedings of the sixth international radiopharmaceutical dosimetry symposium, Oak Ridge, TN, USA, ORISE 00-0164 487-499

Index

S. Mattsson and C. Hoeschen (eds.), *Radiation Protection in Nuclear Medicine*,
DOI 10.1007/978-3-642-31167-3, © Springer-Verlag Berlin Heidelberg 2013

MIX
Papier aus verantwortungsvollen Quellen
Paper from responsible sources
FSC® C105338

If you have any concerns about our products,
you can contact us on
ProductSafety@springernature.com

In case Publisher is established outside the EU,
the EU authorized representative is:
Springer Nature Customer Service Center GmbH
Europaplatz 3, 69115 Heidelberg, Germany

Printed by Libri Plureos GmbH
in Hamburg, Germany